Universe Guide to

Stars and Planets

Ian Ridpath

Illustrated by
Wil Tirion

UNIVERSE BOOKS
New York

Acknowledgements

This book is the result of an unusual international collaboration between an author and an illustrator in different countries, united by their common fascination with the sky. International cooperation of another kind is apparent in the photographic illustrations of this book, which are the result of the efforts of astronomers from many nations working with the world's finest telescopes. These awesome images rank with the most enduring works of art produced by mankind. It is fitting that the observatories concerned have made their results so readily available. The urge to study the sky transcends national boundaries, and so it should be. The skies are open to us all.

The authors would like to thank those whose photographs are used in this book, in particular the following who supplied specific images to our request: David Malin, Anglo-Australian Observatory; Brian Hadley, Royal Observatory Edinburgh; Walter Bonsack, University of Hawaii; the Royal Astronomical Society. Pictures are credited individually where they appear. Dr Peter Adams of the Institute of Geological Sciences provided prints of the U.S. Air Force Lunar Reference Mosaic for making the Moon maps; the US Defense Mapping Agency gave permission for their use.

The following publications have been consulted extensively during the preparation of this work: *Sky Atlas 2000.0* by Wil Tirion (Sky Publishing Corp., Cambridge, Massachusetts and Cambridge University Press, England); *Sky Catalogue 2000.0* edited by Alan Hirshfeld and Roger Sinnott (Sky Publishing Corp. and Cambridge University Press); *Atlas of the Heavens Catalogue* by Antonin Bečvar (Sky Publishing Corp.); *Burnham's Celestial Handbook* by Robert Burnham, Jr. (Dover, New York; Constable, London); *Star Names, Their Lore and Meaning* by R. H. Allen (Dover and Constable); *Astronomy* Magazine (Milwaukee); *Sky & Telescope* Magazine (Cambridge, Mass.).

The Index was skilfully compiled for us by Enid Lake, former librarian of the Royal Astronomical Society. We also acknowledge with gratitude the cheerful and unflagging assistance of our editor Amanda Kent at Collins.

I.R.
W.T.

Published in the United States of America in 1985 by Universe Books
381 Park Avenue South, New York, N.Y. 10016

Printed in Great Britain

Library of Congress Cataloging in Publication Data
Ridpath, Ian.
   Universe guide to stars and planets.
   Includes index.
   1. Astronomy—Amateurs' manuals.   I. Tirion, Wil.
II. Title.
QB63.R53 1985   523   84–24133
ISBN 0–87663–366–1

Contents

Section I

Section II

SECTION I

Introduction

The night sky is one of the most beautiful sights in nature. Yet many people remain lost among the jostling crowd of stars, and are baffled by the progressively changing appearance of the sky from hour to hour and season to season. The charts and descriptions in this book will guide you to the most splendid celestial sights, many of them within the range of simple optical equipment such as binoculars, and all accessible with an average-sized telescope as used by amateur astronomers.

It must be emphasized that you do not need a telescope to take up stargazing. First, use the charts in this book to find your way among the stars with your own eyes and the aid of binoculars, which bring the stars more readily into view. (If you do not already own binoculars they are a worthwhile investment, being relatively cheap, easy to carry and useful for many other purposes than stargazing.)

In the night sky, stars appear to the naked eye as spiky, twinkling lights. Those stars near the horizon seem to flash and change colour. The twinkling and flashing effects are due not to the stars themselves but to the Earth's atmosphere: turbulent air currents cause the star's light to dance around. The steadiness of the atmosphere is referred to as the *seeing*. Steady air means good seeing. The spikiness of star images is due to optical effects in the observer's eye. In reality, stars are spheres of gas similar to our own Sun, emitting their own heat and light.

Stars come in various sizes, from giants to dwarfs, and in a range of colours according to their temperature. At first glance all stars appear white, but more careful inspection reveals that certain stars appear orange-red, notably Betelgeuse, Antares, Aldebaran and Arcturus, while others such as Rigel, Spica and Vega have a bluish tinge. Binoculars bring out the colours more readily than the naked eye. Section II of this book explains more fully the different types of star that exist.

By contrast, planets are cold bodies that shine by reflecting the Sun's light. The planets are constantly on the move as they orbit the Sun, so they cannot be shown on the maps of this book. The planets are described in Section II.

About 2000 stars are visible to the naked eye on a clear, dark night, but you will not need to learn all of them. Start by learning the positions of the brightest stars and major constellations, and use these as signposts to the fainter, less prominent stars and constellations. Once you know the main features of the night sky, you will never again be lost among the stars.

Constellations   The sky is divided into 88 sections, known as constellations, which astronomers use as a convenient way of locating and naming celestial objects. Each constellation is given a separate chart and description in this book. The main constellations of the sky were devised at the dawn of history, by Middle Eastern peoples who fancied that they could see a likeness to certain fabled creatures and mythological heroes among the stars. In particular, the 12 constellations of the zodiac, whose names are familiar to us from the astrological columns in newspapers and magazines, were of importance in the most ancient times. The zodiacal constellations are those that the Sun passes in front of in its yearly path around the heavens. However, it should be realized that the astrological 'signs' of the zodiac are not the same as the modern astronomical constellations, even though they share the same names.

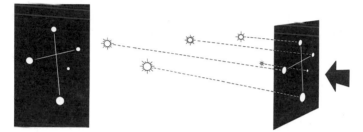

Stars in a constellation are usually unrelated to each other. Here the stars of Crux, the Southern Cross, are shown as they appear from Earth (left) and as they actually lie in space (right). *Wil Tirion.*

Most of the stars in a constellation have no real connection with each other at all; they may all lie at vastly differing distances from Earth, and simply form a pattern by chance. Some of the constellation patterns are easier to recognize than others, such as the magnificent Orion or the distinctive Cassiopeia. Others are faint and obscure, such as Lynx and Telescopium.

Our modern constellations derive from a list of 48, recognized by the Greek astronomer Ptolemy in 150 AD. This list was expanded by navigators and celestial mapmakers, notably the German Johann Bayer (1572–1625), the Pole Johannes Hevelius (1611–1687) and the Frenchman Nicolas Louis de Lacaille (1713–1762). Lacaille introduced 14 new constellations, named after instruments used by scientists and artists, in parts of the southern sky not visible from Mediterranean regions; other astronomers invented constellations to fill in the gaps between the figures recognized by the Greeks. The whole process sounds rather arbitrary, and indeed it was. A number of the newly devised patterns fell into disuse, leaving a total of 88 constellations that were officially adopted by the International Astronomical Union, astronomy's governing body, in 1930.

Star names The main stars in each constellation are labelled with a letter of the Greek alphabet, the brightest star usually (though not always!) being termed α (alpha). Notable exceptions include the constellations Orion and Gemini, in which the stars marked β (beta) are in fact the brightest. (For reference, the entire Greek alphabet is given in the table.) Particularly confusing are the constellations Vela and Puppis, which were once joined with Carina to make the larger figure of Argo Navis, the Ship. As a result of Argo's trisection, neither Vela nor Puppis possess stars labelled α (alpha) or β (beta), and there are gaps in the sequence of Greek letters in Carina.

Greek alphabet:

α	alpha	ι	iota	ϱ	rho
β	beta	κ	kappa	σ	sigma
γ	gamma	λ	lambda	τ	tau
δ	delta	μ	mu	υ	upsilon
ε	epsilon	ν	nu	φ	phi
ζ	zeta	ξ	xi	χ	chi
η	eta	ο	omicron	ψ	psi
θ	theta	π	pi	ω	omega

The system of naming stars by Greek letters was introduced by Johann Bayer, so these designations are often known as Bayer letters. The genitive case of the constellation's name is always used when referring to a star within it; hence Canis Major, for instance, becomes Canis Majoris, and the name α (alpha) Canis Majoris means 'the star α (alpha) in Canis Major'. All constellation names have standard abbreviations. For instance, in abbreviated form Canis Major becomes CMa.

In some constellations, fainter stars are assigned Roman letters, such as L Puppis and P Cygni. An additional system of identifying stars is that of Flamsteed numbers, from their number in a star catalogue drawn up by England's first Astronomer Royal, John Flamsteed (1646–1719). Examples are 61 Cygni and 70 Ophiuchi.

Before the International Astronomical Union's ruling in 1930, there were no officially defined constellation boundaries; stars of one constellation could overlap parts of another constellation. Since 1930 some stars allocated by the Flamsteed system to one constellation have found themselves transferred to a neighbouring constellation, while retaining their old Flamsteed numbers. One example referred to in this book is 41 Lyncis (formerly in Lynx), which under the 1930 boundary rules now lies in Ursa Major.

Prominent stars also have proper names by which they are commonly known. α (alpha) Canis Majoris is better known as Sirius, the brightest star in the sky. Stars' proper names originate from many sources. Some, such as Sirius, Castor and Pollux, date back to Greek times. Many others, such as Aldebaran, are of Arabic origin. Still others were added more recently by European astronomers who borrowed Arabic words in corrupted form; an example is Betelgeuse, which in its current form is meaningless in Arabic. To add to the confusion, spelling of proper names can vary from list to list, and some stars have more than one name.

Star clusters, nebulae and galaxies have a different system of identification. The most prominent of them are given numbers prefixed by the letter M from a catalogue compiled by the Frenchman Charles Messier. M 1 is the Crab Nebula, M 31 the Andromeda Galaxy, and so on. Messier's catalogue contained 103 objects (a few more were added later by other astronomers). A far more comprehensive listing, containing many thousands of objects, is the New General Catalogue (NGC) compiled by J. L. E. Dreyer, with two supplements called the Index Catalogues (IC). Both the Messier numbers (where they exist) and NGC numbers remain in use by astronomers, and both are used in this book.

Star brightnesses Stars appear of different brightnesses in the sky for two reasons. Firstly, they do not all give out the same amount of light. But also, and just as importantly, they all lie at vastly differing distances. Hence, a modest star that is quite close to us can appear brighter than a tremendously powerful star that is a long way away.

Astronomers call a star's brightness its *magnitude*. The magnitude scale was introduced by the Greek astronomer Hipparchus in 129 BC. Hipparchus divided the naked-eye stars into six classes of brightness, from 1st magnitude (the brightest stars) to 6th magnitude (the faintest visible to the naked eye). In his day there was no means of measuring star brightnesses precisely, so this rough classification sufficed. But with the coming of technology it was possible to measure star brightnesses to an exact fraction of a magnitude.

In 1856 the English astronomer Norman Pogson put the magnitude scale on a precise scientific footing, by defining a star of magnitude 1 as being exactly 100 times brighter than a star of magnitude 6. Since, on this scale, a difference of five magnitudes corresponds to a brightness difference of 100 times, a step of one magnitude is equal to a brightness difference of just over 2.5 times (the fifth root of 100).

Objects more than 100 times brighter than 6th magnitude are given negative (minus) magnitudes. For example Sirius, the brightest star in the sky, is of magnitude -1.46. Stars fainter than magnitude 6 are given progressively larger positive magnitudes. The faintest objects detected with telescopes on Earth are of around magnitude 24. The magnitude system may sound confusing at first, but it works well in practice and has the advantage that it can be extended infinitely in both directions, to the very bright and the very faint.

When used without further qualification, the term magnitude refers to the brightness that the star appears in the sky; strictly it should be termed *apparent magnitude*. But because the distance of a star affects the brightness that it appears, the apparent magnitude bears little relation to its actual light output, or *absolute magnitude*. A star's absolute magnitude is defined as the brightness it would appear if it were at a standard distance of 10 parsecs from us (the parsec is explained below); the absolute magnitude is calculated by astronomers from knowledge of the star's nature and its distance.

Absolute magnitude is a good way of comparing the intrinsic brightness of stars. For instance, our daytime star the Sun has an apparent magnitude of -26.8, but an absolute magnitude of 4.8. Deneb (α (alpha) Cygni) has an apparent magnitude of 1.3, but an absolute magnitude of -7.5. From this comparison we deduce that Deneb gives out over 80,000 times as much light as the Sun and is hence one of the most luminous stars known, even though there is nothing at first sight to mark it out as extraordinary.

Magnitudes listed in various catalogues may sometimes not agree with each other; this is because of differences in the instruments used to measure the brightnesses of stars. The values contained in this book come from *Sky Catalogue 2000.0* (Sky Publishing Corporation). In addition, a number of stars actually vary in their light output for various reasons. The nature of such so-called variable stars is discussed in Section II of this book.

Star distances In the Universe, distances are so huge that astronomers have abandoned the tiny kilometre (km) and have invented their own units. Most familiar of these is the *light year* (l.y.), the distance that a beam of light travels in one year. Light moves at the fastest known speed in the Universe, 299,792.5 km per second. A light year is equivalent to 9.46 million million km. On average, stars are several light years apart. For instance, the closest star to the Sun (actually a family of three stars) is 4.3 light years away.

The distances of the nearest stars can be found directly in the following way. The star's position is measured accurately when the Earth is on one side of the Sun, and then remeasured six months later when the Earth has moved around its orbit to the other side of the Sun. When viewed from two widely differing points in space in this way, a nearby star will appear to have shifted slightly in position with respect to more distant stars. This effect is known as *parallax*, and applies to any object viewed from two vantage points against a fixed background, such as a tree against the horizon. A star's parallax shift is so small as to be unnoticeable for all normal purposes – in the case of α (alpha) Centauri, which has the greatest parallax shift of any star, the amount is about the same as the width of a small coin seen at a distance of 2 km. Once

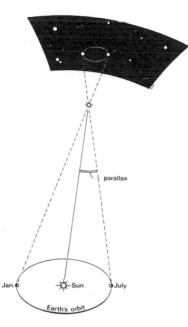

Parallax: as the Earth moves around its orbit, so a nearby star appears to change in position against the celestial background. The shift in position is known as the star's parallax. The nearer the star is to us, the greater its parallax. In this diagram, the amount of parallax is exaggerated for clarity. *Wil Tirion.*

the star's parallax shift has been measured, a simple calculation reveals how far away it is.

An object close enough to us to show a parallax shift of 1″ (one second) of arc would, in the jargon of professional astronomers, be said to lie at a distance of one *parsec*. In practice, no star is this close; the parallax of α (alpha) Centauri is 0.75″ of arc. One parsec is equivalent to 3.26 light years. Astronomers frequently use parsecs in preference to light years because of the ease of converting parallax into distance: a star's distance in parsecs is simply the inverse of its parallax in seconds of arc. For example, a star 2 parsecs away has a parallax of 0.5″ of arc, 4 parsecs away it has a parallax of 0.25″ of arc and so on.

The farther away a star is, the smaller its parallax. Beyond about 50 light years, a star's parallax becomes too small to be measured accurately by telescopes on Earth. Reliable parallaxes have been established for less than 1000 stars. For more distant stars, astronomers first of all estimate the star's absolute magnitude by studying the spectrum of its light. They then compare this estimated absolute magnitude with the observed apparent magnitude to determine the star's distance. The distance obtained in this way is open to considerable error, and the values quoted in various books and catalogues often vary widely because of this.

Distances of stars and galaxies given in this book are all expressed in light years. They are taken from *Sky Catalogue 2000.0*, supplemented where necessary by the *Atlas Coeli Catalogue*. Distances are given to two significant figures. Apart from the closest stars, whose distances have been determined from precise parallax measurements, not too much reliance should be placed on the second digit. Such is the uncertainty in estimating the distance of the more remote stars that in some cases even the first digit may be wrong. Nevertheless it seemed better to give at least a rough figure rather than nothing at all.

Star positions To determine positions of objects in the sky, astronomers use a system of coordinates similar to latitude and longitude on Earth. The celestial equivalent of latitude is called *declination*, and the equivalent of longitude is called *right ascension*. Declination is measured in degrees, minutes and seconds (abbreviated °, ′, and ″) of arc from 0° on the celestial equator to 90° at the celestial poles. The celestial equator is the projection onto the sky of the Earth's equator, and the celestial poles lie exactly above the Earth's poles. Right ascension is measured in hours, minutes and seconds (abbreviated hrs, min and sec), from 0–24 hrs. The 0 hr line of right ascension, the celestial equivalent of the Greenwich meridian, is defined as the point where the Sun crosses the

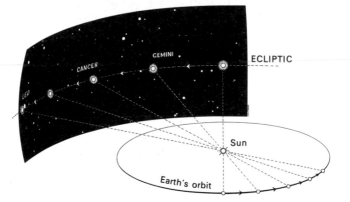

The ecliptic, the Sun's path around the sky each year, is actually a result of the Earth's motion in orbit around the Sun. As the Earth moves along its orbit, the Sun is seen in different directions against the star background. The constellations that the Sun passes in front of during the year are known as the constellations of the zodiac. *Wil Tirion.*

celestial equator on its way north each year. Technically, this point is known as the spring (or vernal) equinox.

The Sun's path around the sky each year is known as the *ecliptic*. This path is inclined at 23.5° to the celestial equator, because the Earth's axis is inclined at 23.5° to the vertical. The most northerly point that the Sun reaches each year is called the *summer solstice*, 23.5° north of the equator, and the most southerly point is the *winter solstice*, 23.5° south of the equator. If the Earth's axis were directly upright with respect to its orbit around the Sun, then the equator and ecliptic would

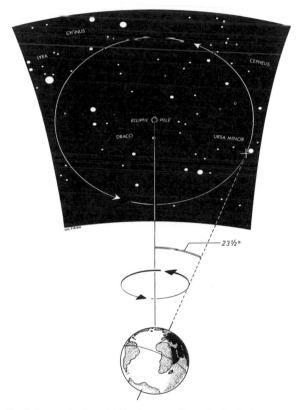

The Earth is very slowly wobbling in space like a tilted spinning top, an effect known as precession. As a result of precession, the position of the celestial poles is constantly changing. The celestial poles trace out a complete circle on the sky every 26,000 years. Only the north celestial pole is shown here, but the effect applies to both poles. *Wil Tirion.*

coincide. One result would be that we would have no seasons on Earth, for the Sun would always remain directly above the equator.

One additional effect that becomes important over long periods of time is that the Earth is slowly wobbling on its axis, like a spinning top. The axis remains inclined at an angle of 23.5°, but the position in the sky to which the north and south poles of the Earth are pointing moves slowly. The Earth's poles describe a large circle on the sky, taking 26,000 years to return to their starting places. Hence the position of the celestial pole is always changing, albeit imperceptibly, as is the position at which the Sun's path (the ecliptic) cuts the celestial equator. This wobbling of the Earth in space is termed *precession*. As an example of the effects caused by precession, whereas Polaris is the Pole Star today, in 12,000 years the north celestial pole will lie near Vega. And the vernal equinox which lay in Aries 2000 years ago now lies in Pisces.

The effect of precession means that the coordinates of celestial objects – the catalogued positions of stars, galaxies, and even constellation boundaries – are continually drifting. Astronomers usually draw up catalogues and star charts for a standard reference date or *epoch*. The epoch of the star charts in this book is the year 2000. For most general purposes, precession does not introduce a noticeable error until after about 50 years, so the charts in this book will be usable without amendment until halfway through the 21st century.

Proper motions   All the stars visible in the sky are members of a vast wheeling mass of stars called the Galaxy. The stars visible to the naked eye are among the nearest to us in the Galaxy. The more distant stars in the Galaxy crowd together in a hazy band called the Milky Way, which can be seen crossing the night sky.

The Sun and the other stars are all orbiting the centre of the Galaxy; it is so huge that the Sun takes about 250 million years to complete one orbit. Other stars move at different speeds, like cars in different lanes on a highway. The result is that stars are all very slowly changing their positions relative to each other. Such stellar movement is termed *proper motion*. Astronomers can detect proper motions with precision measuring techniques, but the motions are undetectable to the naked eye even over a human lifetime. Ancient Greek astronomers catapulted at least 2000 years forward in time would notice little difference in the sky, with the notable exception of Arcturus, a fast-moving bright star, which has moved more than two Moon diameters from its position in Greek times. Over very long periods of time the proper motions of stars considerably distort the shapes of all constellations. (The diagrams show some familiar constellations as they appear now, and as they will appear 100,000 years from now.)

Appearance of the sky   Three factors affect the appearance of the sky: the time of night; the time of year; and your latitude on Earth.

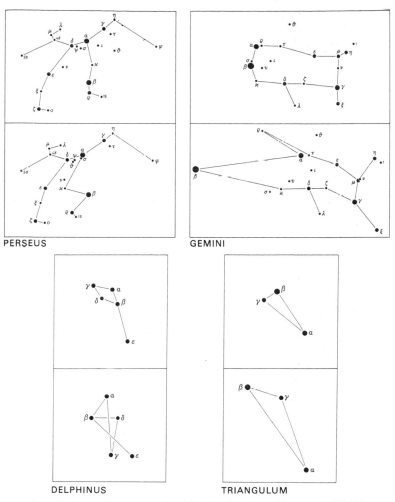

PERSEUS

GEMINI

DELPHINUS

TRIANGULUM

Constellations as they appear today (above) and as they will appear in 100,000 years time (below). *Wil Tirion.*

Firstly, let's consider the effect of latitude. An observer at one of the Earth's poles (latitude 90°) would see the celestial pole directly overhead, and as the Earth turned all stars would circle around the celestial pole without rising or setting. At the other extreme, an observer stationed exactly at the Earth's equator, latitude 0°, would see the celes-

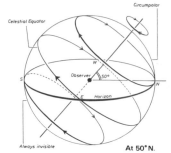

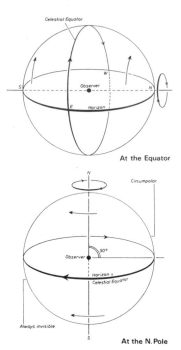

At the Equator

Always invisible   At 50° N.

At the N. Pole

The changing appearance of the sky as seen from different locations on Earth. At the equator, all the sky is visible; as the Earth rotates, stars appear to rise in the east and set in the west. At the pole, by contrast, only one half of the sky is ever visible, the other half being permanently below the horizon. At intermediate latitudes, the situation is between the two extremes. Part of the sky is always above the horizon (the part marked circumpolar), but an equal part is always below the horizon and hence is invisible. Stars between these two regions rise and set during the night. *Wil Tirion*.

tial equator directly overhead. The north and south celestial poles would lie on the north and south horizons respectively, and both hemispheres of the sky would be visible. All stars would rise in the east and set in the west as the Earth rotated.

For most observers, the real situation lies between these two extremes: the celestial pole is at some intermediate altitude between horizon and zenith, and the stars closest to it circle around it without setting (they are said to be *circumpolar*) while the rest of the stars rise and set. The exact angle from the horizon to the celestial pole depends on the observer's latitude. For someone at latitude 40° north, for instance, the north celestial pole is 40° above the northern horizon. If you were at latitude 30° south, the south celestial pole would be 30° above the southern horizon. In other words, the altitude of the celestial pole above the horizon is exactly equal to your latitude, a fact long appreciated by navigators.

As the Earth turns, the stars march across the heavens at the rate of 15° per hour (the Earth rotates through 360° in 24 hours). Therefore the appearance of the sky changes with the time of night. An added complication is that the Earth is also orbiting the Sun each year, so the constellations visible will change with the seasons. For example, a brilliant constellation such as Orion, splendidly seen in December and January, will be in the daytime sky six months later and hence be invisible. The maps in this book will help you to find out which stars are on view whenever and wherever you want to observe.

Using the star charts On pages 16–19 are charts showing the complete northern and southern hemispheres of the sky. In addition to the main stars of each hemisphere, the charts depict the hazy band of the Milky Way; the dotted line is the ecliptic, the Sun's path through the heavens. When the planets are visible, they will be found near to the ecliptic.

Around the rim of each chart are listed the months of the year to help you find which constellations are best placed at about 10 pm local time each month (11 pm when daylight-saving time is in operation). Observers in mid-northern latitudes should take the northern hemisphere chart and turn it so that the month of observation is at the bottom. The chart will show the sky as visible when you face due south that evening. Rotate the chart 15° anticlockwise for each hour after 10 pm, and turn it clockwise for each hour before 10 pm. Observers in mid-southern latitudes should take the southern hemisphere chart and turn it so that the month of observation is at the bottom. The chart will show the stars as they appear when you are facing due north. Turn the chart 15° clockwise for each hour after 10 pm, and 15° anticlockwise for each previous hour. (In all cases above, read 11 pm when daylight-saving time is in operation).

Next comes a series of maps showing the sky as seen when facing north or south at 10 pm (11 pm daylight-saving time) in mid-month from various latitudes. The first set of maps are usable from latitudes 60° north to 10° north; the second set ranges from the equator to 50° south. (They will also be usable for 10° either side of this range without significant error). Curved dotted lines on each map show the horizon from each latitude. Taking these in conjunction with the maps of the complete celestial hemispheres, you should be able to identify the stars in the sky no matter where you are on Earth.

The centrepiece of this book consists of detailed charts and descriptions of each constellation. All stars down to magnitude 5.5 are shown. Some fainter stars have been added in areas of particular interest. The total number of stars shown on these maps is about 3000. All maps are to the same scale, with the exceptions of Hercules, Ursa Major and Serpens, which are slightly smaller in scale, and the rambling constellation Hydra, which is drawn to a significantly smaller scale.

We hope that the charts and descriptions in this book will serve as trusty companions for many nights of exploration under the stars. Good stargazing!

Northern Hemisphere
Stereographic projection

October

September

AQUARIUS

PISCES

PEGASUS

EQUULEUS

AUGUST

AQUILA

ANDROMEDA

DELPHINUS

SAGITTA

LACERTA

VULPECULA

CYGNUS

CEPHEUS

SCUTUM

CASSIOPEIA

SERPENS CAUDA

LYRA

CAMELOPARDALIS

18h

+90°

July

+80°

URSA MINOR

+70°

HERCULES

DRACO

+60°

OPHIUCHUS

URSA MAJOR

+50°

CORONA
BOREALIS

+40°

CANES VENATICI

SERPENS CAPUT

BOÖTES

+30°

COMA BERENICES

June

+20°

LIBRA

+10°

ECLIPTIC

15h

0°

VIRGO

May

−10°

12h April

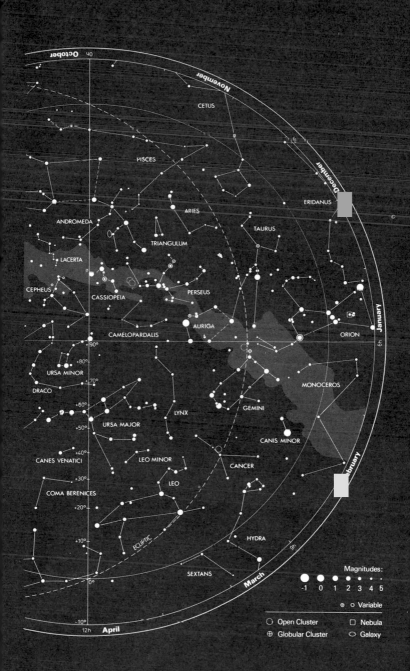

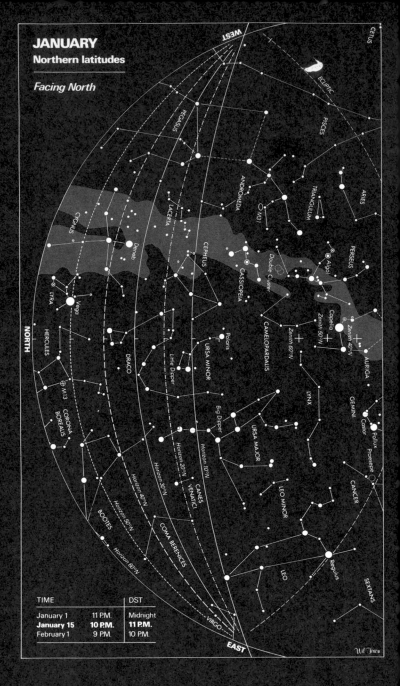

JANUARY
Northern latitudes

Facing North

TIME		DST
January 1	11 P.M.	Midnight
January 15	**10 P.M.**	**11 P.M.**
February 1	9 P.M.	10 P.M.

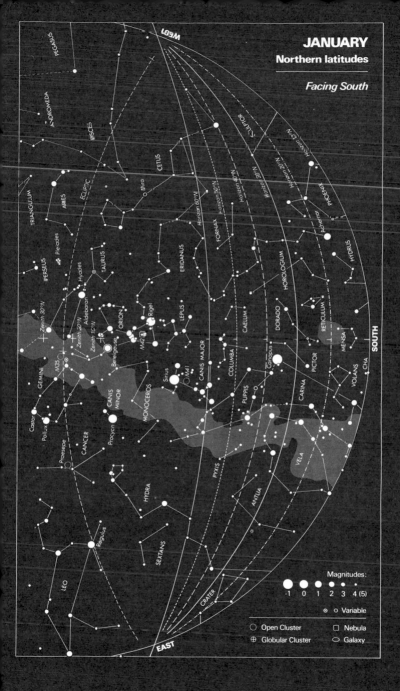

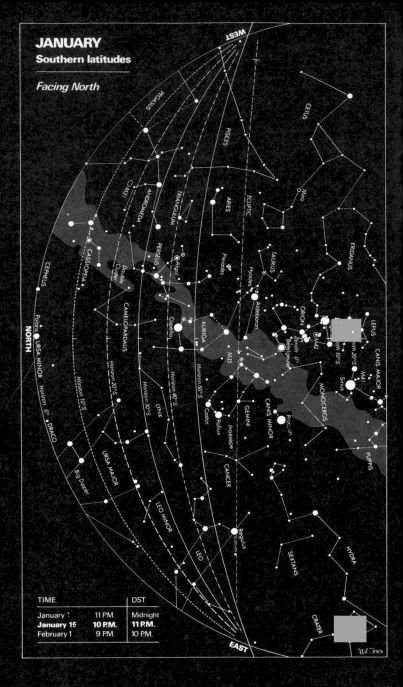

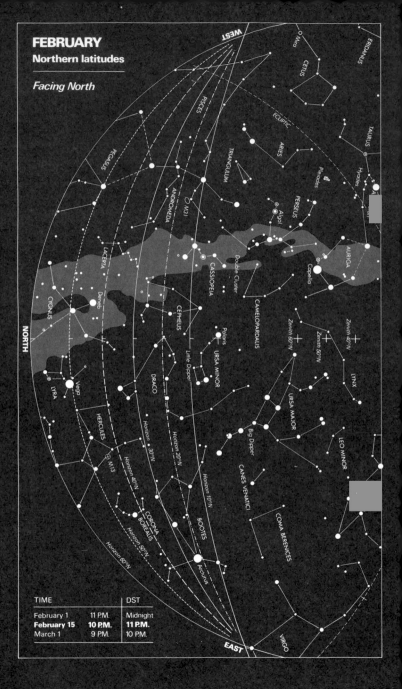

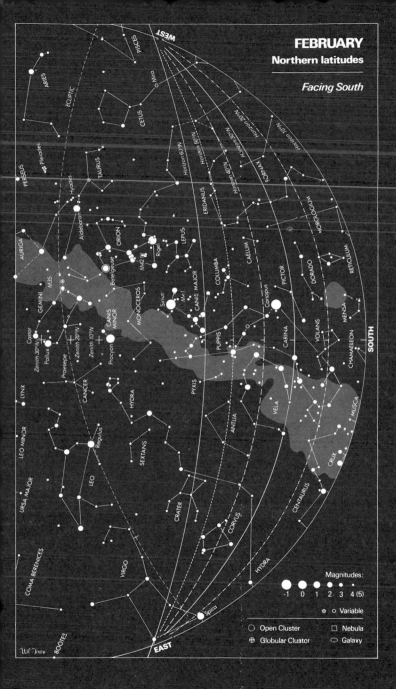

FEBRUARY
Northern latitudes

Facing South

FEBRUARY
Southern latitudes

Facing North

TIME		DST
February 1	11 P.M.	Midnight
February 15	**10 P.M.**	**11 P.M.**
March 1	9 P.M.	10 P.M.

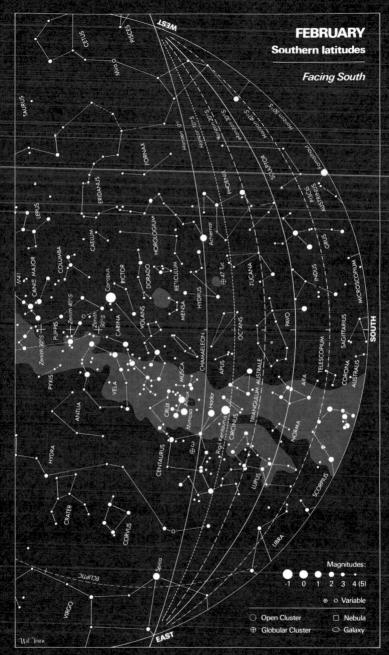

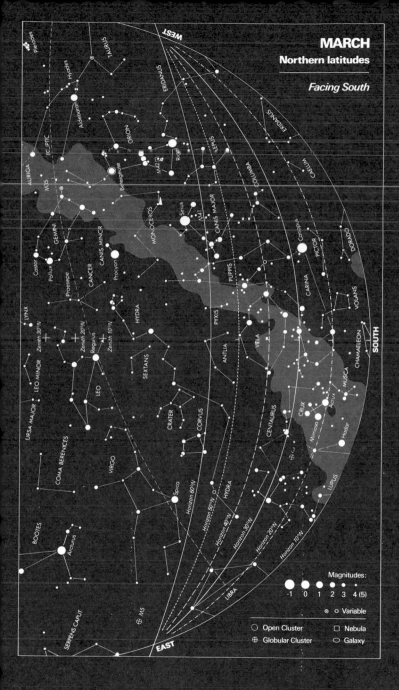

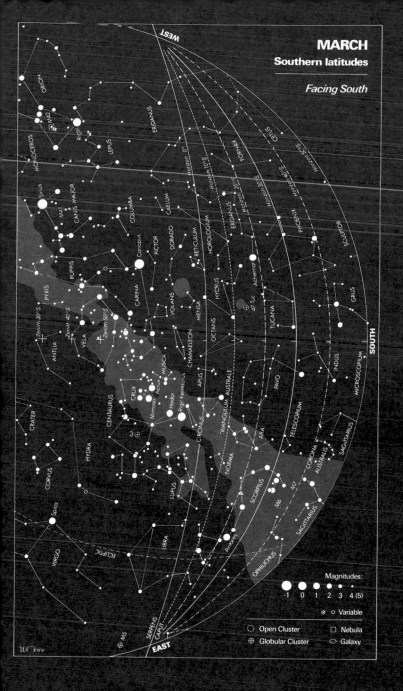

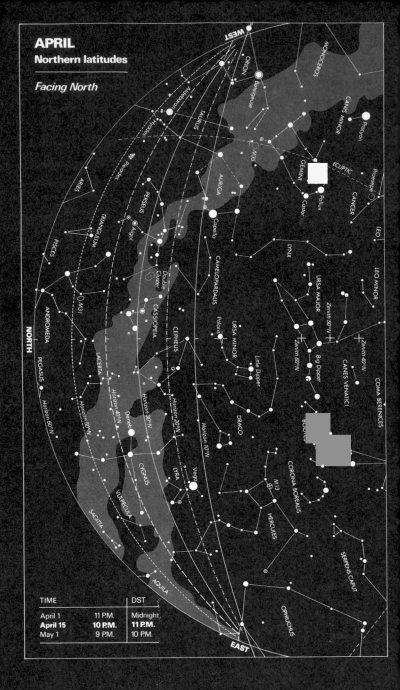

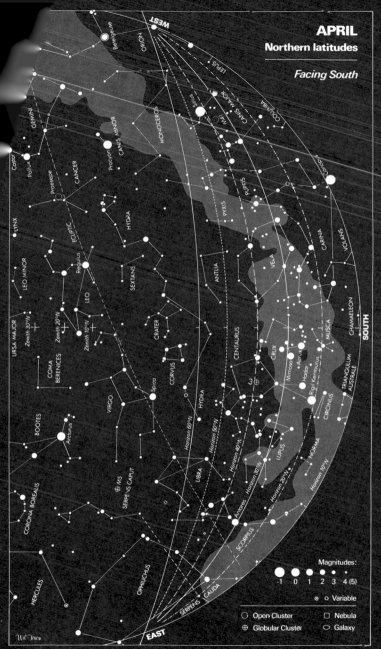

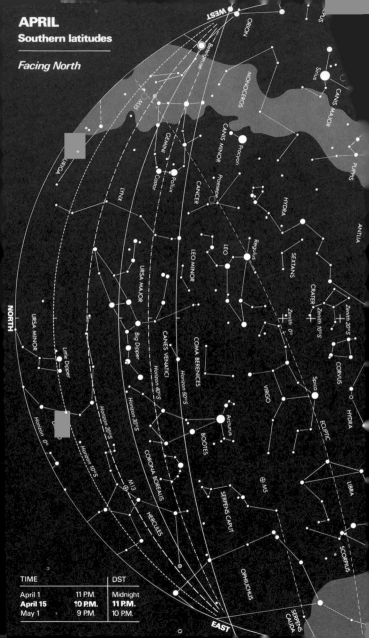

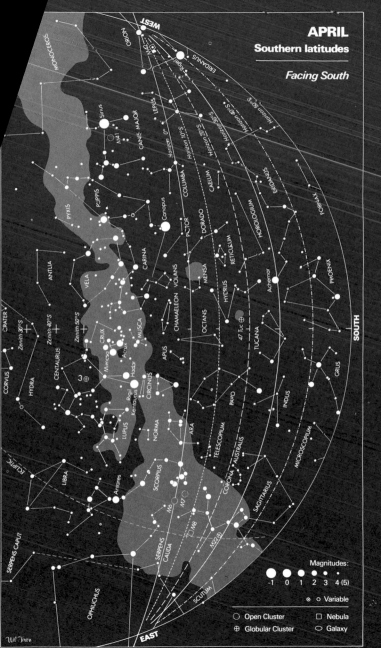

MAY
Northern latitudes

Facing North

TIME		DST	
May 1	11 P.M.	Midnight	
May 15	**10 P.M.**	**11 P.M.**	
June 1	9 P.M.	10 P.M.	

MAY
Northern latitudes

Facing South

Magnitudes:
-1 0 1 2 3 4 (5)

⊙ ○ Variable

○ Open Cluster   □ Nebula
⊕ Globular Cluster   ⬭ Galaxy

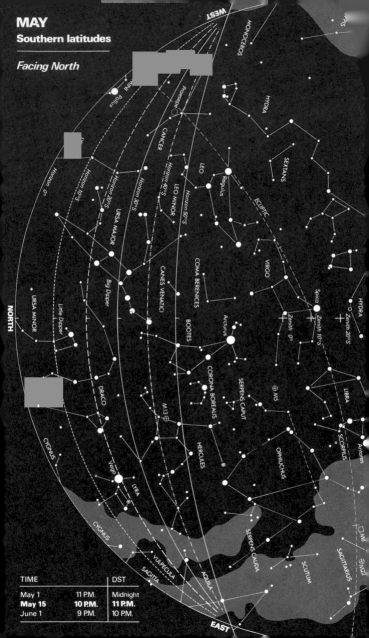

MAY
Southern latitudes

Facing North

TIME		DST
May 1	11 P.M.	Midnight
May 15	**10 P.M.**	**11 P.M.**
June 1	9 P.M.	10 P.M.

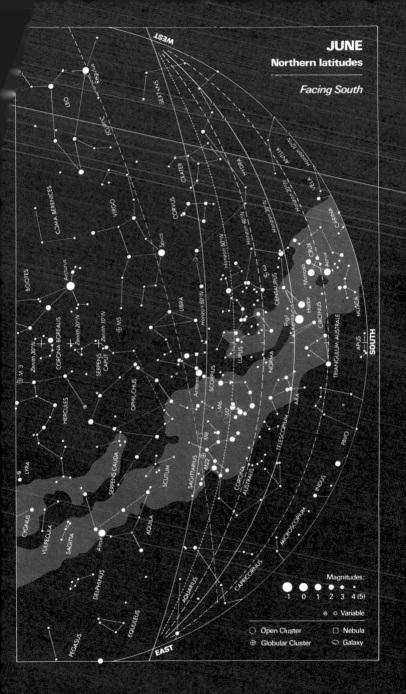

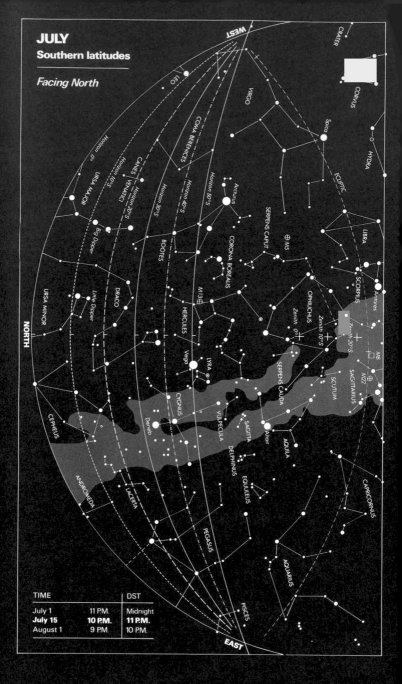

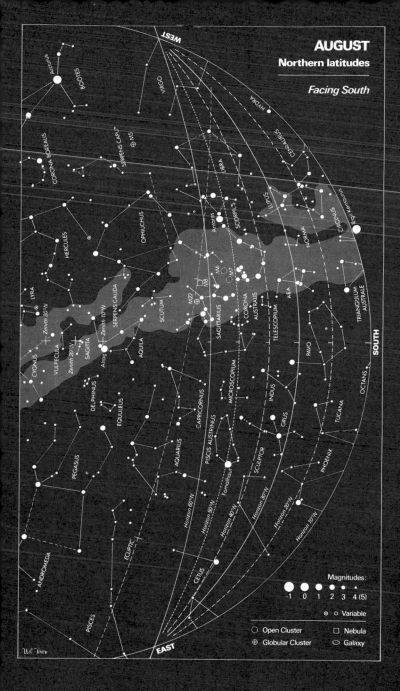

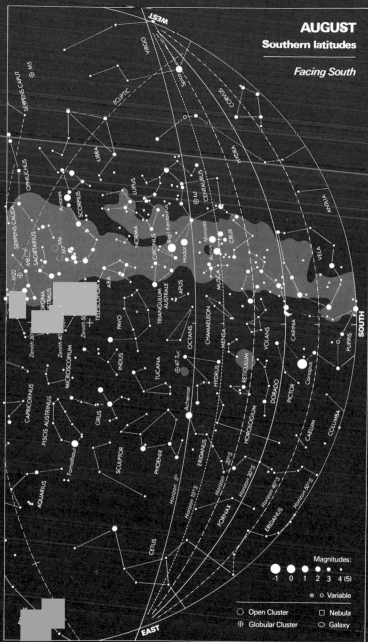

AUGUST
Southern latitudes

Facing South

WEST

VIRGO
SERPENS CAUDA
OPHIUCHUS
⊕ M5
ECLIPTIC
SERPENS CAPUT
LIBRA
CORVUS
HYDRA
LUPUS
⊕ ω CENTAURUS
Antares
SCORPIUS
NORMA
Rigil Kentaurus
ANTLIA
CIRCINUS
Mimosa
CRUX
VELA
Hadar
Acrux
MUSCA
X8 ⊕ M7 M6
SAGITTARIUS
ARA
APUS
CHAMAELEON
CARINA
M22 ⊕
CORONA AUSTRALIS
TELESCOPIUM
TRIANGULUM AUSTRALE
MENSA
VOLANS
Zenith 30°
Zenith 40°
Zenith 50°
PAVO
OCTANS
⊕ 47 Tuc
HYDRUS
RETICULUM
DORADO
PICTOR
Canopus
SOUTH
INDUS
TUCANA
Achernar
HOROLOGIUM
MICROSCOPIUM
CAPRICORNUS
GRUS
PHOENIX
ERIDANUS
CAELUM
COLUMBA
PISCIS AUSTRINUS
SCULPTOR
20°S
Fomalhaut
AQUARIUS
CETUS
Horizon 0°
Horizon 10°S
Horizon
FORNAX
Horizon 30°S
ERIDANUS
Horizon 40°S
Horizon 50°S

Magnitudes:
-1 0 1 2 3 4 (5)

⊙ ○ Variable

○ Open Cluster   □ Nebula
⊕ Globular Cluster   ○ Galaxy

EAST

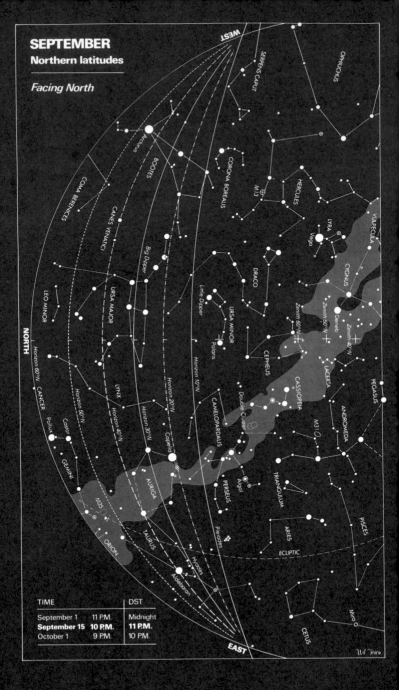

SEPTEMBER
Northern latitudes

Facing North

TIME		DST	
September 1	11 P.M.	Midnight	
September 15	**10 P.M.**	**11 P.M.**	
October 1	9 P.M.	10 P.M.	

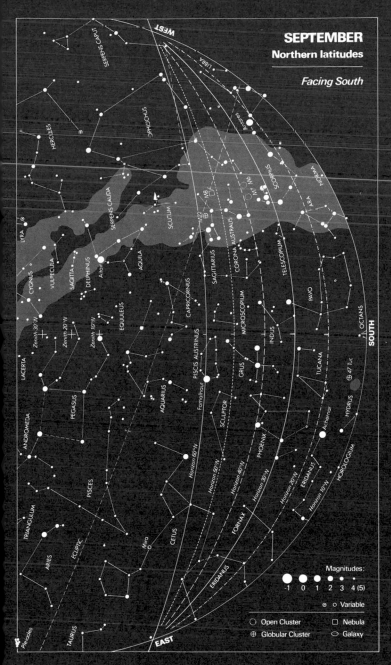

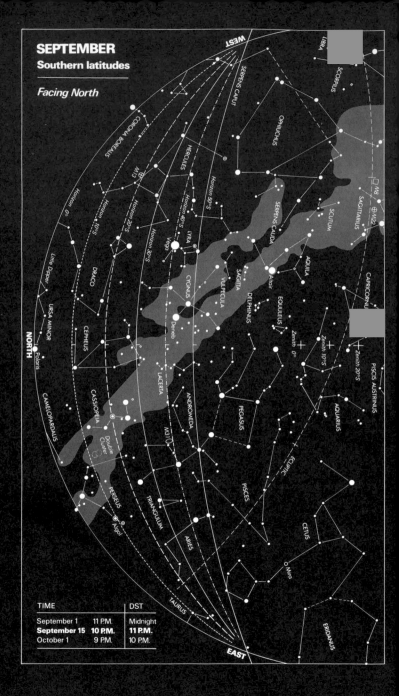

SEPTEMBER
Southern latitudes

Facing North

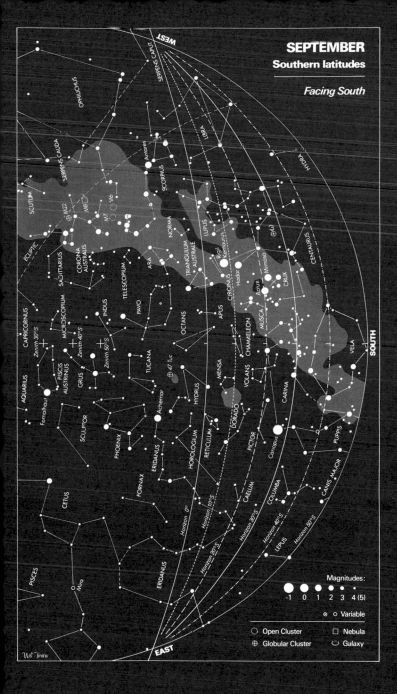

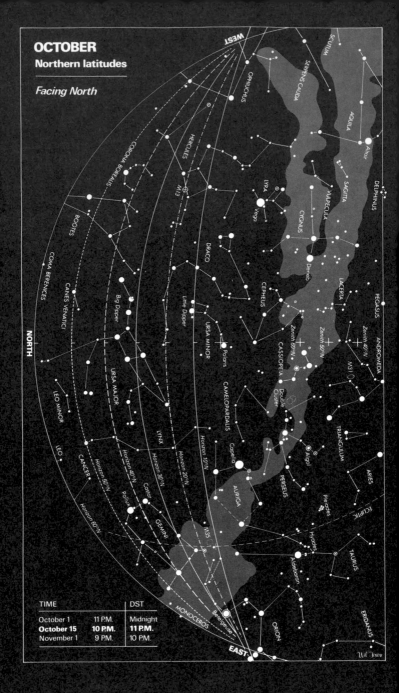

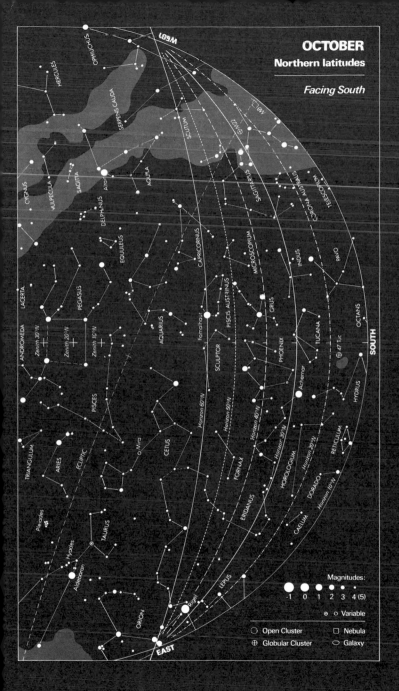

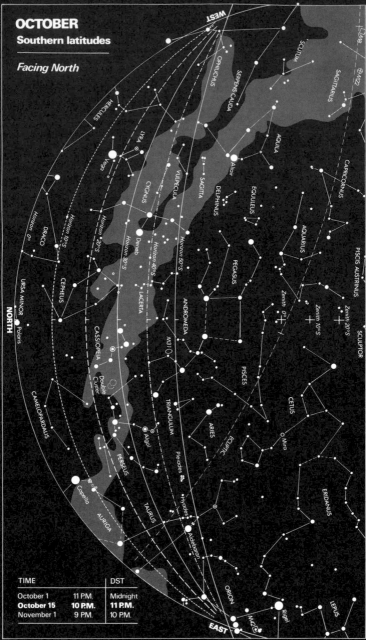

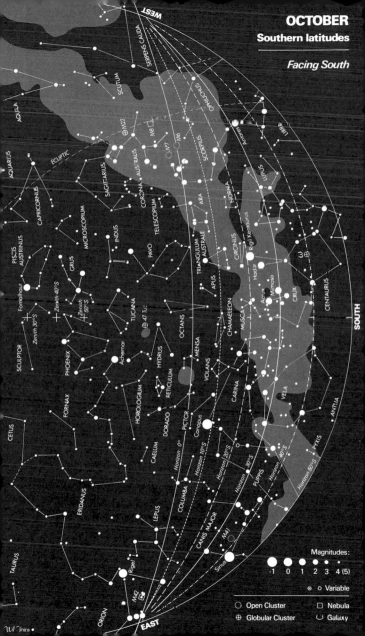

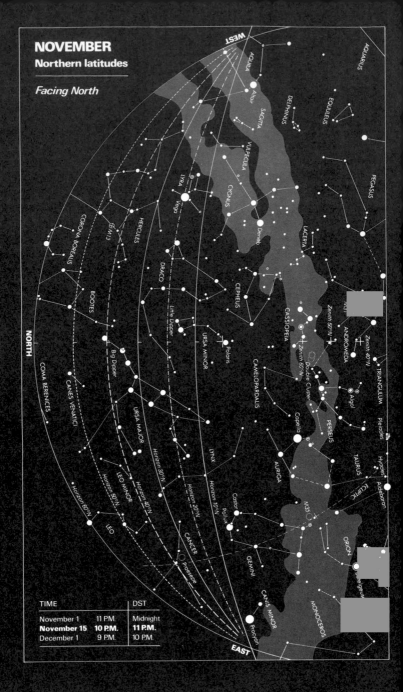

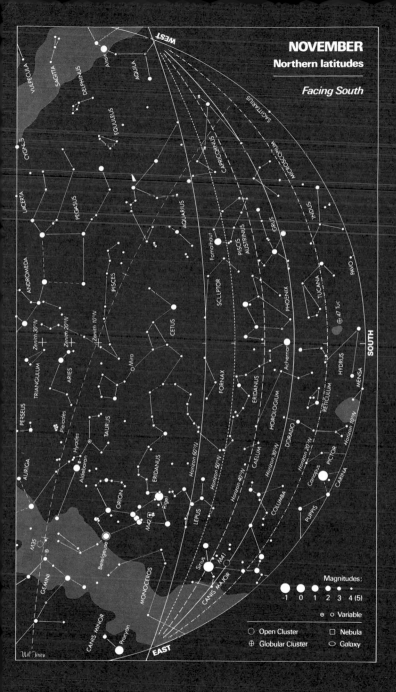

NOVEMBER
Northern latitudes

Facing South

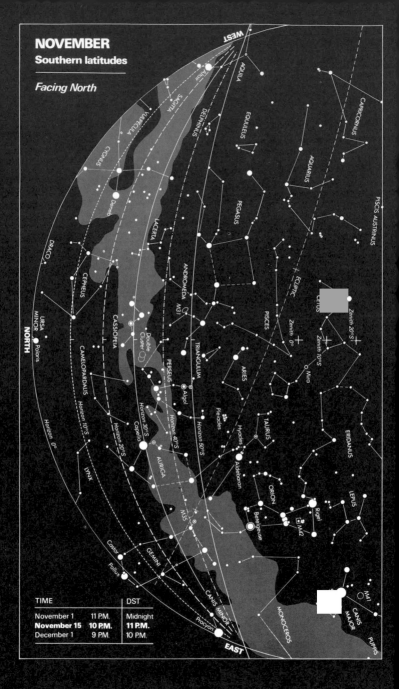

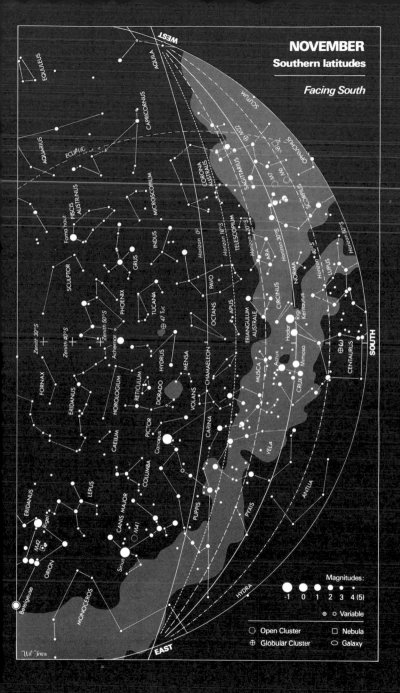

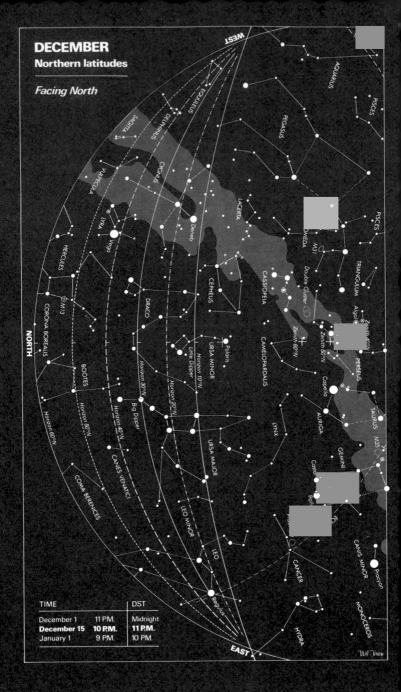

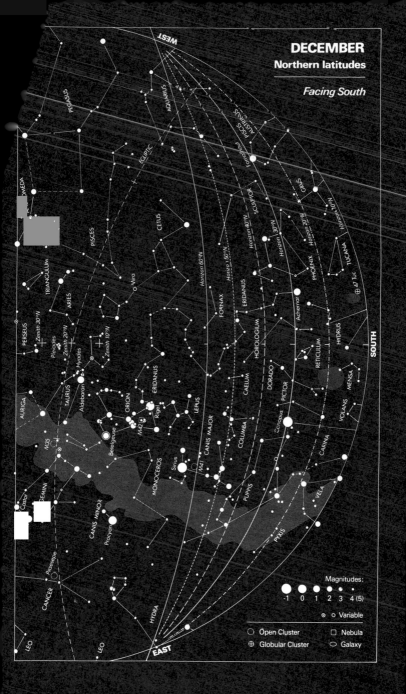

DECEMBER
Northern latitudes

Facing South

WEST

SOUTH

EAST

PEGASUS
AQUARIUS
PISCIS AUSTRINUS
GRUS
SCULPTOR
TUCANA
ANDROMEDA
PISCES
CETUS
PHOENIX
ECLIPTIC
Fomalhaut
47 Tuc
Horizon 60°N
Horizon 50°N
Horizon 40°N
Horizon 30°N
Horizon 20°N
Horizon 10°N
TRIANGULUM
ARIES
o Mira
FORNAX
ERIDANUS
HOROLOGIUM
Achernar
HYDRUS
RETICULUM
PERSEUS
Zenith 30°N
Pleiades
Zenith 20°N
Hyades
Zenith 10°N
CAELUM
DORADO
MENSA
TAURUS
Aldebaran
ORION
ERIDANUS
LEPUS
PICTOR
VOLANS
AURIGA
M42
Rigel
CANIS MAJOR
COLUMBA
Canopus
CARINA
M35
Betelgeuse
Sirius
M41
PUPPIS
MONOCEROS
VELA
Castor
GEMINI
CANIS MINOR
Procyon
PYXIS
Praesepe
CANCER
HYDRA
LEO
LEO

Magnitudes:
-1 0 1 2 3 4 (5)

⊛ ○ Variable

○ Open Cluster   □ Nebula
⊕ Globular Cluster   ⬭ Galaxy

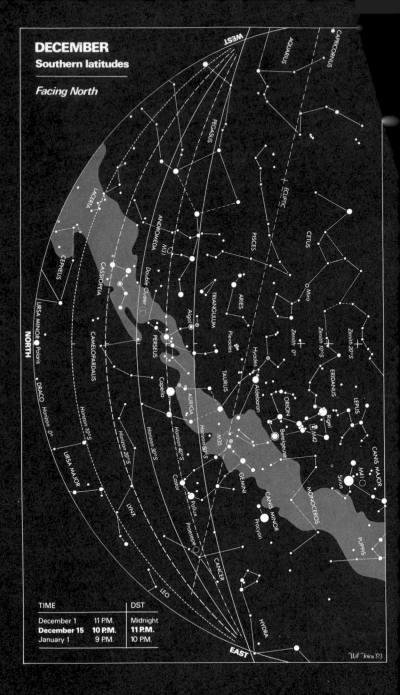

ANDROMEDA

Andromeda represents the daughter of Queen Cassiopeia who ▮ chained to a rock as a sacrifice to the sea monster Cetus until saved ▮ Perseus, whom she subsequently married. The constellation originate ▮ in ancient times. Despite its fame, Andromeda is not particularly strik-ing: its brightest star is of only 2nd magnitude. Its most prominent feature is a line of four stars extending from the Square of Pegasus; one of these stars marks a corner of the Square, although it is actually part of Andromeda. This star, known both as Sirrah and Alpheratz, marks the head of the chained Andromeda; another star in the line, Mirach, represents her waist, and a third, Alamak, is her chained foot. The most celebrated object in the constellation is the Andromeda Galaxy, M 31, a spiral galaxy like our Milky Way; it is the most distant object visible to the naked eye. Two stars leading from Mirach (β (beta) Andromedae) act as a guide to it.

α (alpha) Andromedae (Sirrah or Alpheratz), mag. 2.1, is a blue-white star 105 l.y. away.

β (beta) And (Mirach), mag. 2.1, is a red giant 88 l.y. away.

γ (gamma) And (Alamak or Almach), 160 l.y. away, is an outstanding triple star. Its two brightest components, of mags. 2.2 and 5.0, form one of the finest pairs for small telescopes: their colours are yellow and blue. The fainter, blue star also has a close 6th mag. blue companion, requiring apertures of 220 mm or over to resolve.

δ (delta) And, mag. 3.3, is an orange giant 160 l.y. away.

μ (mu) And, mag. 3.9, is a white star 82 l.y. away.

π (pi) And, 390 l.y. away, is a blue-white star of mag. 4.4 with a mag. 8.7 com-panion visible in small telescopes. ▶

The Andromeda Galaxy, M 31, and its two small companion galaxies, M 32 just below centre, and the larger but fainter NGC 205 to its upper right. *Hale Observatories photograph.*

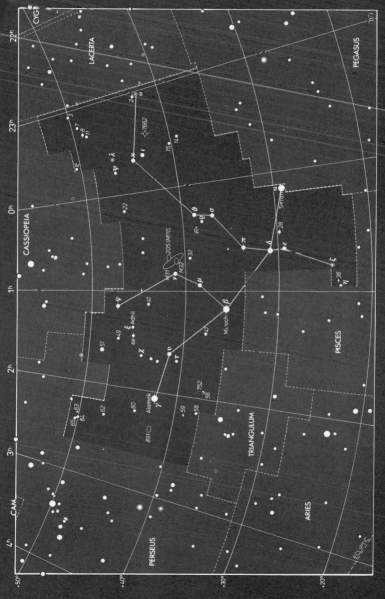

ANTLIA   The Air Pump

An obscure southern hemisphere constellation introduced in 1763 by the French astronomer Nicolas Louis de Lacaille to commemorate the air pump invented by the English physicist Robert Boyle. Lacaille, the first man to map the southern skies completely (which he did from an observatory at the Cape of Good Hope), introduced several new constellations to fill gaps in between existing constellations. Most of these new figures are unremarkable, as is Antlia.

α (alpha) Antliae, mag. 4.3, the constellation's brightest star, is an orange giant 280 l.y. distant.

δ (delta) Ant, 1100 l.y. away, is a blue-white star of mag. 5.6 with a mag. 9.7 companion visible in small telescopes.

$\zeta^1 \zeta^2$ (zeta^1 zeta^2) Ant is a wide pair of stars of mags. 5.8 and 5.9 visible in binoculars. Small telescopes show that ζ^1 (zeta^1) Ant is itself double, with components of mags. 6.4 and 7.2.

NGC 3132 is a relatively large and bright planetary nebula, of 8th mag. and appearing larger in the telescope than Jupiter. Its central star is of 10th mag. It lies 1300 l.y. away, on the border with Vela.

◄ 56 And is a binocular duo of 6th mag. yellow giant stars, 240 l.y. away. They are easily split in binoculars; they lie close to the star cluster NGC 752.

M 31 (NGC 224), the Andromeda Galaxy, is a spiral galaxy 2.2 million l.y. away. It is visible to the naked eye as an elliptical fuzzy patch, and becomes more prominent in binoculars or a telescope with low magnification (too high a power reduces the contrast and renders the fainter parts of the galaxy less visible). Dark lanes can be seen in the spiral arms surrounding the nucleus. But the full extent of the galaxy becomes apparent only on long exposure photographs; visual observers see just the brightest, central portion of the galaxy. If the entire Andromeda Galaxy were bright enough to be seen by the naked eye, it would appear five or six times the diameter of the full Moon. M 31 is accompanied by two small satellite galaxies, the equivalent of our Magellanic Clouds. The brighter of these, M32 (NGC 221), is visible in small telescopes as a fuzzy star-like glow 0.5° south of M 31's core. The second companion, NGC 205, is fainter, though larger, and over 1° northwest of M 31.

NGC 752 is a widespread cluster of up to 100 stars, best seen in a telescope, lying 3400 l.y. away.

NGC 7662 is one of the brightest and easiest planetary nebulae to see with a small telescope. At low powers it appears as a fuzzy, 9th mag. blue-green star, but magnifications of × 100 or so reveal its slightly elliptical disk. Larger apertures show a central hole; the central star is a difficult object for amateur telescopes. NGC 7662 lies 1800 l.y. away.

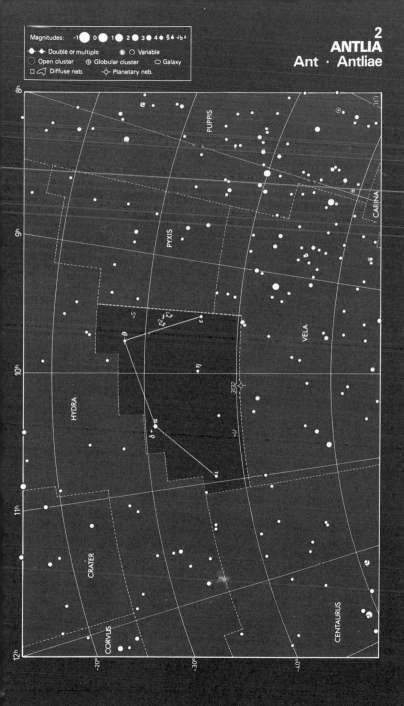

APUS   The Bird of Paradise

A faint constellation near the south celestial pole, introduced in 1603 by the German astronomer Johann Bayer on his *Uranometria*, the first star atlas to cover the entire sky. Bayer introduced a total of 12 new constellations to accommodate the otherwise unattached stars of the southern sky; they were not actually invented by him, but were adapted from the ideas of various seafarers, notably the Dutchman Petrus Theodorus.

α (alpha) Apodis, mag. 3.8, is an orange giant star 220 l.y. away.

β (beta) Aps, mag. 4.2, is a yellow star 110 l.y. away.

γ (gamma) Aps, mag. 3.9, is a yellow star 46 l.y. away.

δ^1 δ^2 (delta^1 delta^2) Aps, 390 l.y. away, is a naked-eye or binocular pair of orange giant stars of mags. 4.7 and 5.3.

NGC 2997 is a handsome 11th mag. galaxy in Antlia, similar to our own Milky Way. *Anglo-Australian Telescope Board.*

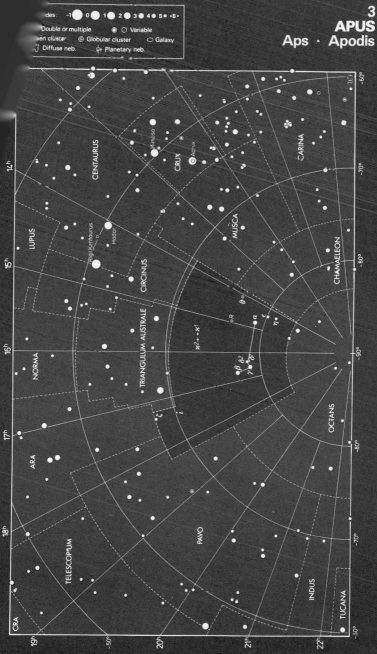

Magnitudes: −1 0 1 2 3 4 5 <5

○ Double or multiple   ◉ ○ Variable
○ Open cluster   ⊕ Globular cluster   ○ Galaxy
Diffuse neb.   ⬥ Planetary neb.

AQUARIUS   The Water Carrier

Aquarius is one of the most ancient constellations. The Babyloni
saw in this area of sky the figure of a man pouring water from a jar, a.
it is still represented this way today. The most prominent part of tl
constellation is the Y-shaped figure of four stars representing the wate.
jar itself, centred on the star ζ (zeta) Aquarii. Aquarius is in an area of
'watery' constellations (including Pisces, Cetus, and Capricornus):
possibly this association with water arose because the Sun passes
through this region during the rainy season. The Sun is in the constellation of Aquarius from late February to early March. Aquarius will one
day contain the vernal equinox, the point at which the Sun crosses into
the northern celestial hemisphere each year. This astronomically
important point, from which the coordinate of right ascension is
measured, will move into Aquarius from neighbouring Pisces, because
of the effect of precession, in about 600 years time. Therefore the so-
called Age of Aquarius, much heralded by astrologers, is a long way
off yet.

Three main meteor showers radiate from Aquarius each year. The
first, the η (eta) Aquarids, is the richest, reaching a maximum of 40
meteors per hour on May 5. The δ (delta) Aquarids, around July 28,
produce about 20 meteors per hour. A weaker stream, the ι (iota)
Aquarids, produces a maximum of 8 meteors per hour on August 6.
Each shower is named after the bright star closest to its radiant.

α (alpha) Aquarii (Sadalmelik, from the Arabic meaning 'lucky one of the king'),
mag. 3.0, is a yellow supergiant 950 l.y. away.

β (beta) Aqr (Sadalsuud, meaning 'luckiest of the lucky'), mag. 2.9, is a yellow
supergiant 980 l.y. away.

γ (gamma) Aqr (Sadachbia), mag. 3.8, is a white star 91 l.y. away.

δ (delta) Aqr (Skat), mag. 3.3, is a giant white star 98 l.y. away.

ε (epsilon) Aqr (Albali), mag. 3.8, is a white star 110 l.y. away.

ζ (zeta) Aqr, 76 l.y. away, is a celebrated binary consisting of twin white stars of
mags. 4.5 and 4.3 orbiting each other every 856 years. At present the two stars
are at their closest together as seen from Earth, and need at least 75 mm aperture
with high magnification to split them.

M 2 (NGC 7089) is a globular cluster easily visible in binoculars or small telescopes, but requiring 250 mm aperture to be resolved into individual stars. M2
is a rich and highly concentrated globular cluster. It lies about 50,000 l.y. away.

M 72 (NGC 6981) is a globular cluster 60,000 l.y. away, much smaller and less
impressive than M 2. It is too faint to be seen in binoculars.

NGC 7009, 1450 l.y. away, is a famous planetary nebula known as the Saturn
Nebula because of its resemblance to that planet when seen in large telescopes. ▶

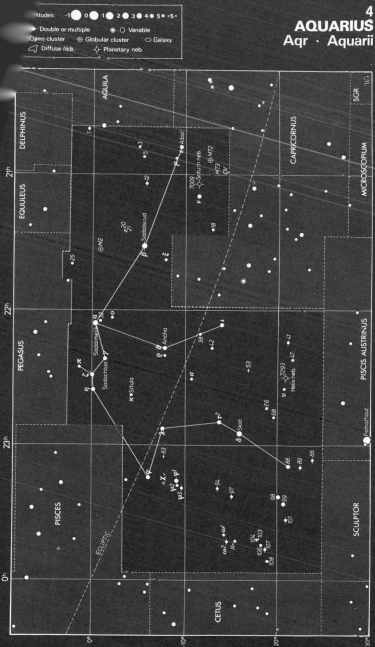

Magnitudes: −1 0 1 2 3 4 5 <5
Double or multiple Variable
Open cluster ⊕ Globular cluster ◯ Galaxy
Diffuse neb. ⬦ Planetary neb.

AQUILA
DELPHINUS
SGR
CAPRICORNUS
MICROSCOPIUM

21h

3
5
Albali
Saturn neb.
M72 ⊕
M73 ⊕
ν
7009

42
ν

20
21
Sadalsuud
18

⊕M2
β
ξ

25
⊙

EQUULEUS

22h

α
ο
Sadalmelik
Ancha
ι
ρ σ
38
42
τ
41
47

PEGASUS
ζ π
η
γ
Sadachbia
κ Situla
53
ν
7293
Helix neb.

σ
66
68

τ²
Skat
δ

PISCIS AUSTRINUS
Fomalhaut

23h
λ
83
88
69
86

χ
ψ¹
ψ²
ψ³
94
97
98
99

PISCES
101

ω¹
R
104
106
103
107
108

ω²

ECLIPTIC

0h

CETUS
SCULPTOR

0°
−10°
−20°
−30°

AQUILA   The Eagle

A constellation dating from ancient times, representing the bird that mythology was the companion of Jupiter, and often carried his thunder bolts. Aquila's brightest star, Altair, forms one corner of the Summer Triangle that is completed by Deneb in Cygnus and Vega in Lyra. Altair is easily identified by the two fainter stars β(beta) and γ (gamma) Aquilae which stand like sentinels either side of it. These stars are called Alshain and Tarazed, both from the Persian name for the constellation, Shahin tara zed, the Star-striking Falcon. Aquila lies in the Milky Way and contains rich star fields, particularly towards the neighbouring constellation Scutum. Aquila is an abundant area for novae.

α (alpha) Aquilae (Altair, an Arabic name meaning flying eagle), mag. 0.77, is a white star 16 l.y. away, among the Sun's closest neighbours.

β (beta) Aql (Alshain), mag. 3.7, is a yellow star 42 l.y. away.

γ (gamma) Aql (Tarazed), mag. 2.7, is a yellow giant star 280 l.y. away.

δ (delta) Aql, mag. 3.4, is a white star 52 l.y. away.

η (eta) Aql, 1400 l.y. away, is one of the brightest Cepheid variable stars. Its brightness ranges from mag. 4.1 to mag. 5.3 every 7.2 days.

15 Aql, 390 l.y. away, is a yellow giant star of mag. 5.4 with a purplish mag. 7.2 companion easily visible in small telescopes.

57 Aql, 590 l.y. away, is an easy double for small telescopes, consisting of a bluish star of mag. 5.7 with a mag. 6.5 companion.

NGC 6709 is a loosely scattered cluster of about 40 stars of mags 9–11, approximately 2500 l.y. away.

◄ But in most amateur telescopes, of 75 mm aperture or more, it appears as an 8th mag. blue-green ellipse. It has a 12th mag. central star.

NGC 7293, 690 l.y. distant, is the nearest planetary nebula to the Sun, and is commonly known as the Helix Nebula. It is the largest planetary nebula in apparent size, covering 0.25° of sky, half the apparent size of the Moon. Despite its size, the Helix Nebula appears quite faint, and is best found with binoculars or very low power on a telescope, when it appears as a circular misty patch, not as impressive as its large size would suggest.

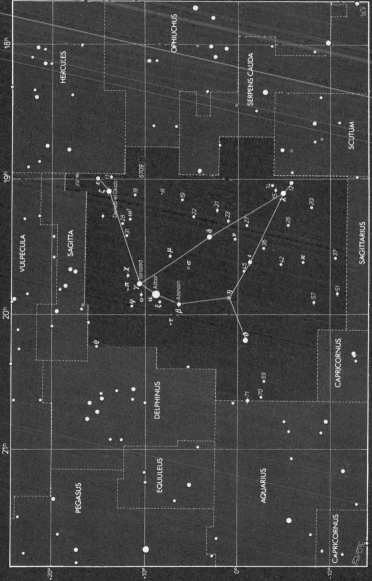

ARA   The Altar

This constellation, although relatively faint and little-known, originated in Greek and Roman times, when it was visualized as the altar on which Centaurus, the Centaur, was about to sacrifice Lupus, the Wolf. Another legend saw it as the Altar of the Gods. Ara lies in a rich part of the Milky Way, south of Scorpius.

α (alpha) Arae, mag. 3.0, is a blue-white star 220 l.y. away.

β (beta) Ara, mag. 2.9, is a yellow supergiant 780 l.y. away.

γ (gamma) Ara, mag. 3.3, is a blue-white giant star 1100 l.y. away.

δ (delta) Ara, mag. 3.6, is a blue-white star 150 l.y. away.

ζ (zeta) Ara, mag. 3.1, is an orange giant star 140 l.y. away.

NGC 6193 is a bright cluster of about 30 stars 2300 light years away, the brightest member of which is mag. 5.6. Associated with the cluster is an irregular patch of nebulosity, NGC 6188.

NGC 6397 is a large and bright globular cluster easily visible with binoculars or a small telescope. It is among the closest globular clusters to us, 7500 l.y. away.

The stars of the 7th mag. globular cluster NGC 6397 in Ara appear more widely scattered than in many other globulars. *Anglo-Australian Telescope Board.*

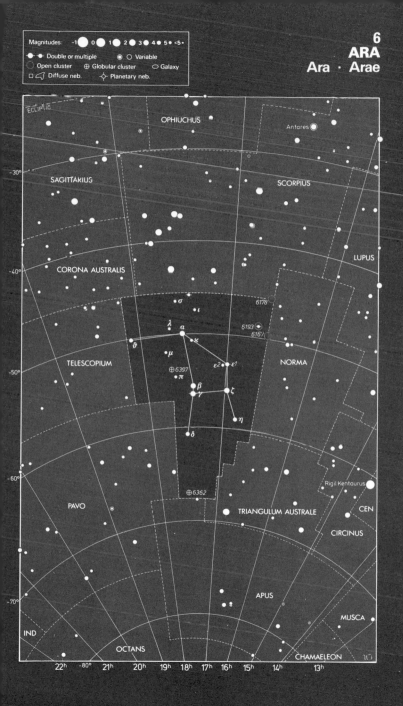

Magnitudes: −1 ● 0 ● 1 ● 2 ● 3 ● 4 ● 5 ● <5 ·

●—● Double or multiple   ◉ ○ Variable
○ Open cluster   ⊕ Globular cluster   ○ Galaxy
□ ⌐ Diffuse neb.   ◇ Planetary neb.

ECLIPTIC

OPHIUCHUS

Antares ◉

−30°

SAGITTARIUS

SCORPIUS

LUPUS

−40°

CORONA AUSTRALIS

6178

σ

ι

λ   α

5193
616'/

ϑ

×

TELESCOPIUM

μ

ε² ● ε¹

NORMA

⊕ 6397

−50°

π

β

γ   ζ

η

δ

⊕ 6362

−60°

Rigil Kentaurus

PAVO

TRIANGULUM AUSTRALE

CEN

CIRCINUS

−70°

APUS

MUSCA

IND

OCTANS

CHAMAELEON

22ʰ   −80°   21ʰ   20ʰ   19ʰ   18ʰ   17ʰ   16ʰ   15ʰ   14ʰ   13ʰ

ARIES The Ram

A constellation whose origin dates back to ancient times, lying between Taurus and Andromeda. In legend, Aries represented the Ram whose golden fleece was sought by Jason and the Argonauts. The Arabs knew it as the Sheep, and the Chinese as a Dog. Despite its faintness, Aries has assumed great importance in astronomy, because 2000 or so years ago it contained the point where the Sun passed from south to north across the celestial equator each year. This point, the vernal equinox, marked the start of northern hemisphere spring, and from it the celestial coordinate known as right ascension is measured; this point was also known as the First Point of Aries. Because of the slight wobble of the Earth in space known as precession, this point has now moved into Pisces (see diagram), but for historical reasons the vernal equinox is still sometimes referred to as the First Point of Aries.

α (alpha) Arietis (Hamal, from the Arabic for sheep), mag. 2.0, is a yellow giant star 85 l.y. away.

β (beta) Ari (Sheratan, the sign), mag. 2.6, is a white star 46 l.y. away.

γ (gamma) Ari (Mesarthim) is a striking double star consisting of twin white stars each of mag. 4.6, clearly visible through small telescopes even under low magnification. The stars lie about 160 l.y. away.

ε (epsilon) Ari, 410 l.y. away, is a challenging double star for apertures of 100 mm or over. High magnification reveals a tight pair of white stars of mags. 5.3 and 5.6.

λ (lambda) Ari, 140 l.y. away, is a white star of mag. 4.8 with a yellow 7th mag. companion, easily visible in small telescopes or even good binoculars.

π (pi) Ari, 620 l.y. away, is a blue-white star of mag. 5.2, with a close mag. 8.3 companion, difficult to distinguish in the smallest telescopes.

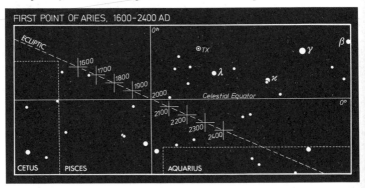

Movement of the so-called First Point of Aries over 800 years. The First Point of Aries currently lies in Pisces and is approaching Aquarius. *Wil Tirion.*

Magnitudes: -1 0 1 2 3 4 5 <5
●—● Double or multiple   ◉ ○ Variable
◌ Open cluster   ⊕ Globular cluster   ◯ Galaxy
▢ Diffuse neb.   ✧ Planetary neb.

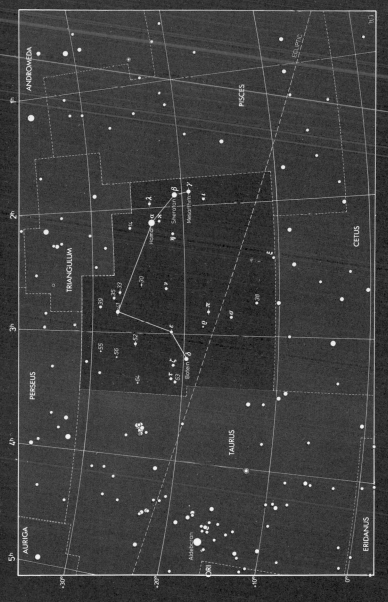

AURIGA   The Charioteer

A large and prominent constellation, known since ancient Babylonian and Greek times as a Charioteer. Auriga's leading star is Capella, 7th-brightest in the whole sky, marking the Charioteer's left shoulder. The Greeks identified Capella with the she-goat Amalthea who helped feed the infant Zeus. The stars ζ (zeta) and η (eta) Aurigae are referred to as the Kids. The star γ (gamma) Aurigae is shared with Taurus the Bull, one of whose horns it marks, and is dealt with under that constellation as β (beta) Tauri.

α (alpha) Aurigae (Capella, little she-goat), mag. 0.08, lies 42 l.y. away. It is actually a spectroscopic binary, consisting of two yellow stars orbiting every 104 days, although they do not eclipse.

β (beta) Aur (Menkalinan, shoulder of the charioteer), 72 l.y. away, is a blue-white eclipsing variable star of approximate mag. 2.0, which varies by about 0.1 of a mag. every 3.96 days.

ε (epsilon) Aur, a white supergiant about 2000 l.y. away, is an eclipsing binary with an exceptionally long period. Normally it shines at mag. 3.0, but every 27 years it sinks to mag. 3.8 as it is eclipsed by a dark companion, remaining fully eclipsed for a year. One theory is that the companion of ε (epsilon) Aurigae is a young star surrounded by a disk of material. Its last eclipse was in 1983.

ζ (zeta) Aur, 520 l.y. away, is an amazing eclipsing binary, consisting of an orange giant orbited every 972 days by a small blue companion. During eclipses, ζ (zeta) Aurigae's brightness drops from mag. 5.0 to 5.7.

θ (theta) Aur, 120 l.y. away, is a blue-white star of mag. 2.6. It has a white companion of mag. 7.5 which, because of its closeness and relative faintness, needs at least 100 mm aperture and high magnification to distinguish. This is a tough double for steady nights.

ω (omega) Aur, 225 l.y. away, is a double star of mags. 4.9 and 8.0, visible in small telescopes.

14 Aur, 105 l.y. away, is a double star of mags. 5.0 and 8.1, split in small telescopes.

UU Aur is a deep red variable star more than 3000 l.y. away. It varies semiregularly between 5th and 7th mags. with a rough period of 235 days.

M 36 (NGC 1960) is a small, bright cluster of about 60 stars, visible in binoculars and resolvable into stars in small telescopes. M 36 lies 3800 l.y. away.

M 37 (NGC 2099) is the richest of the clusters in Auriga, containing about 150 stars. The cluster appears as a hazy, unresolved patch in binoculars, but a 100 mm telescope resolves it into a sparkling field of faint stardust, featuring a brighter orange star at the centre.

M 38 (NGC 1912) is a large, scattered cluster of about 100 faint stars, visible in binoculars, with a noticeable cross-shape when seen through telescopes. Its distance is 3600 l.y. Next to it lies the small fuzzy blob of NGC 1907, a much smaller and fainter cluster.

NGC 2281 is a binocular cluster of about 30 stars, 5400 l.y. away. In a telescope the stars are arranged in a crescent, four brighter stars forming a diamond shape.

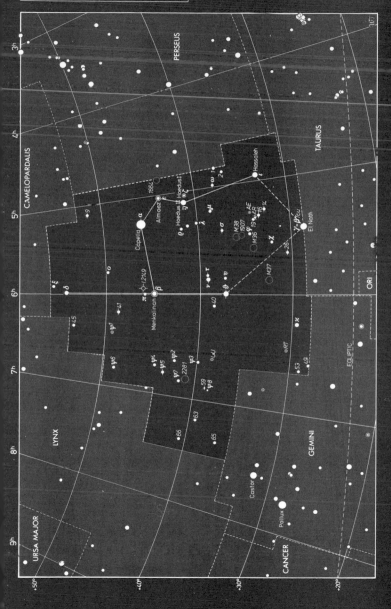

Magnitudes: -1 0 1 2 3 4 5 •<5•
● ● Double or multiple ◉ ○ Variable
◯ Open cluster ⊕ Globular cluster ◯ Galaxy
▢ ◿ Diffuse neb. ◇ Planetary neb.

BOÖTES   The Herdsman

An ancient constellation, representing a Herdsman driving a Bear
(Ursa Major) around the sky; he is often depicted holding the leash of
the Hunting Dogs, Canes Venatici. The name of the constellation's
brightest star, Arcturus, actually means 'bearkeeper' in Greek. Arcturus
is the brightest star in the northern hemisphere of the sky, and is easily
identified: the curving handle of the Big Dipper or Plough acts as a
pointer to it. Arcturus forms the base of a large 'Y' shape with ε (epsilon)
Boötis, γ (gamma) Boötis, and α (alpha) Coronae Borealis. The rest of
the constellation is much fainter than Arcturus, but contains numerous
double stars of interest. The year's most abundant meteor shower, the
Quadrantids, radiate from the northern part of Boötes, an area of sky
that was once occupied by the now-abandoned constellation of Quadrans Muralis, the Mural Quadrant (hence the shower's name). The
Quadrantid meteors reach a peak of about 100 meteors per hour on
January 3–4 each year, although they are not as bright as other rich
showers such as the Perseids.

α (alpha) Boötis (Arcturus), mag. −0.04, is the 4th brightest star in the entire
sky. It is a red giant, 27× the diameter of the Sun, lying 36 l.y. away; its orange-
red colour is noticeable to the naked eye, and is more striking with optical aid.
Arcturus has a mass similar to that of the Sun, and it is believed that our Sun will
swell up to become a red giant like Arcturus in 5000 million years.

β (beta) Boo (Nekkar, corrupted from the Arabic meaning herdsman), mag. 3.5,
is a yellow giant 140 l.y. away.

γ (gamma) Boo (Haris or Seginus), mag. 3.0, is a white star 105 l.y. away.

δ (delta) Boo, mag. 3.5, is a yellow giant 140 l.y. away. It has a wide binocular
companion of mag. 7.5.

ε (epsilon) Boo (Izar, girdle), 150 l.y. away, is a celebrated double star: an orange
giant primary of mag. 2.7 with a blue companion of mag. 5.1. This close double
of contrasting colours requires a telescope of at least 75 mm at × 100 power or
more, because the bright primary tends to overwhelm its fainter companion; but
its appearance when split has led to the alternative name Pulcherrima, meaning
'most beautiful'.

ι (iota) Boo, 91 l.y. away, is a wide double star of mags. 4.8 and 8.3.

κ (kappa) Boo, 125 l.y. away, is an easy double star for small telescopes, consisting
of components of mags. 4.6 and 6.6.

μ (mu) Boo (Alkalurops, staff or crook), 59 l.y. distant, is an attractive triple star.
To the naked eye it appears as a blue-white star of 4th mag., but binoculars reveal
a wide companion. Telescopes of 75 mm aperture with high magnification show
that the companion is itself double, comprising two close 7th mag. stars; these
orbit each other every 260 years.   ▶

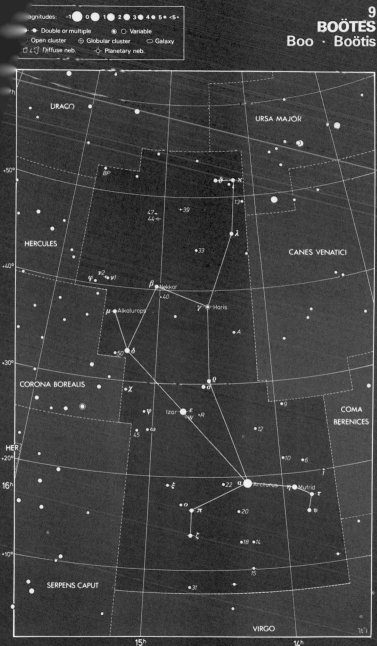

Magnitudes: –1 0 1 2 3 4 5 <5

● Double or multiple ◉ ○ Variable
Open cluster ⊕ Globular cluster ◯ Galaxy
Diffuse neb. Planetary neb.

DRACO

URSA MAJOR

+50°

BP

θ κ

13

47 39
44

λ

HERCULES

33

CANES VENATICI

+40°

ν2
φ ν1

β Nekkar
40
μ Alkalurops

γ Haris

A

δ
50

CORONA BOREALIS

χ

ρ
σ

9

COMA BERENICES

ψ ε Izar
W R

12

+20°

45 ω

10 6
j

HER
16ʰ

ξ

22 α Arcturus
η Mufrid
τ
υ

ο π
20

ζ

18 14

+10°

15

SERPENS CAPUT

31

VIRGO

15ʰ 14ʰ

CAELUM   The Chisel

An obscure, almost irrelevant constellation at the foot of Eridanus, introduced in the 1750s by Nicolas Louis de Lacaille during his mapping of the southern sky.

α (alpha) Caeli, mag. 4.5, is a white star 65 l.y. away.

β (beta) Cae, mag. 5.1, is a white star 55 l.y. away.

γ (gamma) Cae, mag. 4.6, is an orange star 170 l.y. away. It has a mag. 8.5 companion visible in telescopes of 60 mm aperture.

δ (delta) Cae, mag. 5.1, is a blue-white star 750 l.y. away.

◄ ν (nu) Boo is a binocular duo consisting of a white star, mag. 5.0, 170 l.y. distant, and an unrelated orange giant of mag. 5.0, 330 l.y. away.

π (pi) Boo, 360 l.y. away, is a double star with blue-white components of mags. 4.9 and 5.9, visible in small telescopes.

ξ (xi) Boo, 22 l.y. away, is a showpiece double for small telescopes, consisting of yellow and orange stars of mags. 4.8 and 6.9, orbiting each other every 150 years.

Magnitudes: −1 ● 0 ● 1 ● 2 ● 3 ● 4 ● 5 ● <5 •
●—● Double or multiple   ◉ ○ Variable
○ Open cluster   ⊕ Globular cluster   ○ Galaxy
▢ ◠ Diffuse neb.   ◇ Planetary neb.

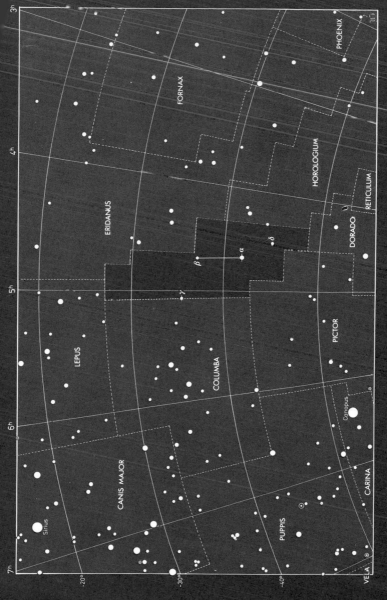

CAMELOPARDALIS   The Giraffe

A faint and obscure constellation, also spelled Camelopardus. It was apparently invented by the Dutch astronomer Petrus Plancius, but first came to prominence in 1624 through a book by the German mathematician Jakob Bartsch, a son-in-law of Johannes Kepler.

α (alpha) Camelopardalis, mag. 4.3, is a blue supergiant star 2800 l.y. away.

β (beta) Cam, at mag. 4.0 the brightest star in the constellation, is a yellow supergiant 1500 l.y. distant. It has a wide 9th mag. companion star visible in small telescopes or even good binoculars.

Σ 1694 (Struve 1694), 400 l.y. away, is a pair of mags. 5.3 and 5.8 white stars easily split in small telescopes.

NGC 1502 is a small star cluster of about 15 members visible in binoculars or small telescopes, somewhat triangular in shape and with two easy double stars at its centre. It lies 3750 l.y. away.

NGC 2403 is a 9th mag. spiral galaxy nearly 0.33° long, easily seen as an elliptical glow in 100 mm telescopes on a good night. It lies at a distance of 11 million l.y.

NGC 2523 is a curiously shaped barred spiral galaxy in Camelopardalis. At 13th mag. it is beyond the range of amateur telescopes. *Hale Observatories photograph.*

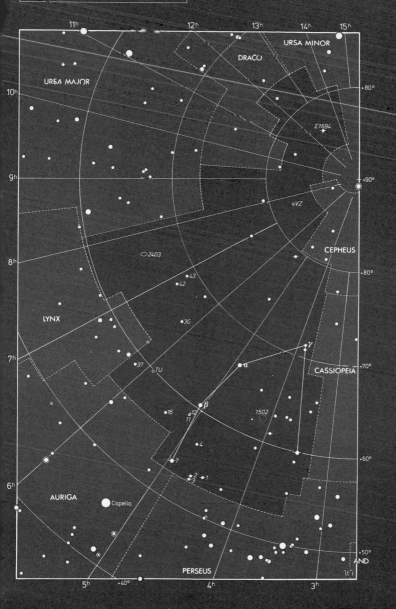

Magnitudes: -1 0 1 2 3 4 5 <5

● Double or multiple   ◉ ○ Variable

○ Open cluster   ⊕ Globular cluster   ○ Galaxy

□ Diffuse neb.   ◇ Planetary neb.

URSA MINOR

DRACO

URSA MAJOR

Σ1694

VZ

2403

CEPHEUS

43

42

3G

LYNX

37   TU

γ

α

CASSIOPEIA

16   β

12   1502

11

4

7

2   1

3

AURIGA   Capella

PERSEUS   AND

CANCER   The Crab

Cancer represents the crab which attacked Hercules when he was fighting with the Hydra; the luckless Crab was crushed underfoot by mighty Hercules, but was subsequently elevated to the heavens. In ancient times, the Sun reached its most northerly point in the sky each year while it was in Cancer. The date on which the Sun is farthest north of the equator, June 21, is known as the summer solstice. On the summer solstice, the Sun appears overhead at noon at latitude 23.5° north on Earth. This latitude came to be known as the Tropic of Cancer, a name it retains today even though, because of the effect of precession, the Sun now lies in the neighbouring constellation of Gemini on the summer solstice. Cancer is the faintest of the 12 constellations of the zodiac through which the Sun passes each year, but it nevertheless contains much of interest, notably the star cluster Praesepe, the Manger. Praesepe is flanked by two stars, Asellus Borealis and Asellus Australis (meaning the northern donkey and the southern donkey, evidently visualized as feeding at the stellar Manger).

α (alpha) Cancris (Acubens, meaning claws), mag. 4.3, is a white star 100 l.y. away. It has an 11th mag. companion visible with telescopes of 75 mm aperture and over.

β (beta) Cnc, mag. 3.5, an orange giant 170 l.y. away, is the brightest star in the constellation.

γ (gamma) Cnc (Asellus Borealis), mag. 4.7, is a white star 230 l.y. away.

δ (delta) Cnc (Asellus Australis), mag. 3.9, is a yellow giant star 220 l.y. away.

ζ (zeta) Cnc, 52 l.y. away, is an interesting multiple star. A small telescope reveals two yellow stars of mags. 5.1 and 6.0. Larger telescopes also reveal that the brighter component has a close 6th mag. companion, which orbits it every 60 years. At their closest, in 1990, the stars need at least 220 mm aperture to separate them, but at their widest 100 mm should split them.

ι (iota) Cnc, 420 l.y. away, is a yellow giant of mag. 4.0 with a blue-white mag. 6.6 companion just visible in binoculars, or easily seen through a small telescope.

M 44 (NGC 2632), Praesepe, commonly called the Beehive Cluster, is a swarm of about 75 stars visible as a misty patch to the naked eye, and best seen in binoculars. Praesepe covers 1.5° of sky, 3× the apparent diameter of the Moon. It lies 520 l.y. away.

M 67 (NGC 2682) is a denser and smaller cluster of stars than the Beehive, visible as a misty patch in binoculars or small telescopes, and needing apertures of at least 75 mm to resolve it into about 60 individual stars. It lies 2700 l.y. away.

12
CANCER
Cnc · Cancri

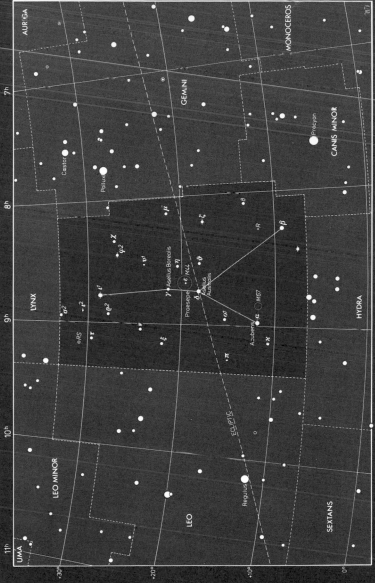

CANES VENATICI   The Hunting Dogs

A constellation introduced in 1690 by the Polish astronomer Johannes Hevelius, consisting of a sprinkling of faint stars below Ursa Major. It represents the dogs Asterion and Chara, held on a leash by Boötes as they pursue the Great Bear around the pole. Canes Venatici contains numerous galaxies, the most famous being M 51, the Whirlpool, a beautiful face-on spiral pictured on page 282; it was the first galaxy in which spiral form was detected, by Lord Rosse, with his 72 inch (1.8 m) reflector at Birr Castle, Ireland, in 1845.

α (alpha) Canum Venaticorum is popularly called Cor Caroli, meaning Charles's heart, a reference to the martyred King Charles I of England; it is said to have shone particularly brightly in 1660 on the arrival of Charles II in England at the Restoration of the Monarchy. It is a double star, of mag. 2.9 with a mag. 5.5 companion easily seen in small telescopes. Both stars are blue-white, but various observers have reported subtle shades of colour when seen through a telescope. The brighter star is the standard example of a rare class of stars with strong and variable magnetic fields. Cor Caroli is 130 l.y. away.

β (beta) CVn (Asterion or Chara), mag. 4.3, is the only other star of any prominence in the constellation. It is a yellow star, 30 l.y. away.

Y CVn is a deep red variable star sometimes known as La Superba. It varies between mags. 5 and 6.5 approximately every 160 days.

M 3 (NGC 5272) is a rich globular cluster located midway between Cor Caroli and Arcturus in Boötes, and is regarded as one of the finest globular clusters in the northern sky. It is beyond naked eye reach, but can be picked up easily as a hazy star in binoculars or a small telescope; a 5th mag. star nearby acts as a guide. In small telescopes M 3 appears as a condensed ball of light with a faint outer halo. Apertures of 100 mm or more are needed to resolve individual stars in its outer regions. M 3 is 45,000 l.y. away.

M 51 (NGC 5194), the Whirlpool Galaxy, is a spiral galaxy 14 million l.y. away with a smaller satellite galaxy lying at the end of one of its arms. It is disappointing in small telescopes, which show a faint milky radiance around the starlike nuclei of the galaxy and its satellite, but it is nevertheless worth hunting for on clear, dark nights.

M 94 (NGC 4736) is a compact spiral galaxy presented nearly face-on. In amateur telescopes it appears like an 8th mag. comet, with a fuzzy starlike nucleus surrounded by an elliptical halo. M 94 is 14 million l.y. away.

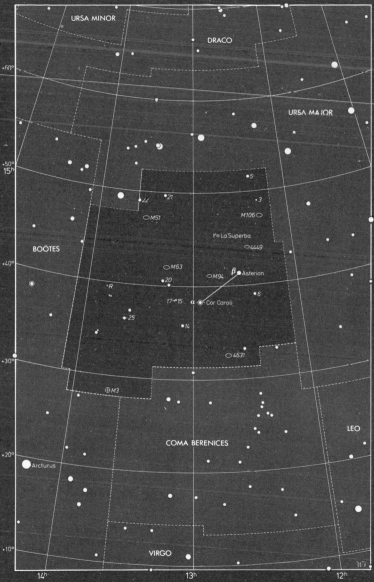

Magnitudes: -1 0 1 2 3 4 5 <5
Double or multiple ○ Variable
Open cluster ⊕ Globular cluster ○ Galaxy
□ Diffuse neb. ◇ Planetary neb.

URSA MINOR

DRACO

URSA MAJOR

+60°

15h

+50°

5

22   21

3

M51

M106

γ La Superba

4449

BOÖTES

+40°

M63

20

β Asterion

M94

6

R

17 15

α Cor Caroli

25

14

+30°

4631

⊕ M3

COMA BERENICES

LEO

+20°

Arcturus

+10°

VIRGO

14h   13h   12h

CANIS MAJOR   The Greater Dog

An ancient constellation, representing one of the two dogs (the other being Canis Minor) following at the heels of Orion. Canis Major contains many brilliant stars, making it one of the most prominent constellations; its leading star, Sirius, is the brightest star in the entire sky. Sirius features in many legends. The ancient Egyptians based their calendar on its yearly motion around the sky.

α (alpha) Canis Majoris (Sirius, from the Greek for sparkling or scorching), mag. −1.46, is a brilliant white star 8.7 l.y. away, one of the Sun's closest neighbours. It has a white dwarf companion of mag. 8.5 that orbits it every 50 years. The brilliance of Sirius overpowers this white dwarf so that even when the two stars are at their greatest separation, as in 1975, telescopes of 200 mm aperture or more and steady atmospheric conditions are required for it to be visible. During the 1990s, when the two stars are at their closest, the companion will be impossible to see in any amateur telescope.

β (beta) CMa (Mirzam, the announcer), mag. 2.0, is a blue giant 720 l.y. away. It is a pulsating star whose variations are undetectable to the naked eye. It changes in brightness by a few hundredths of a mag. every 6 hours.

δ (delta) CMa (Wezen, meaning weight), mag. 1.9, is a yellow supergiant star 3000 l.y. away.

ε (epsilon) CMa (Adhara, the virgins), mag. 1.5, is a blue giant 490 l.y. away. It has an 8th mag. companion which is difficult to see in small telescopes because of the glare from the primary.

η (eta) CMa (Aludra), mag. 2.4, is a blue supergiant 2500 l.y. away.

M 41 (NGC 2287) is a large and bright cluster of about 50 stars, easily visible in binoculars or small telescopes, and detectable by the naked eye under good conditions – it was known to the ancient Greeks. A low-power view in small telescopes shows the individual stars grouped in bunches and curves, covering an area of sky equivalent to the apparent diameter of the Moon. The brightest star in the cluster is a 7th mag. orange giant. M 41 is 2500 l.y. away.

NGC 2362 is a smaller and fainter cluster than M 41, but is easily found surrounding the mag. 4.4 blue giant star τ (tau) Canis Majoris. Small telescopes show about 40 stars in the cluster. τ (tau) CMa is a member of the cluster, which lies about 4000 l.y. away.

Magnitudes: -1 0 1 2 3 4 5 <5
● ━ ● Double or multiple   ◉ ○ Variable
○ Open cluster   ⊕ Globular cluster   ◯ Galaxy
▢ ⌇ Diffuse neb.   ◇ Planetary neb.

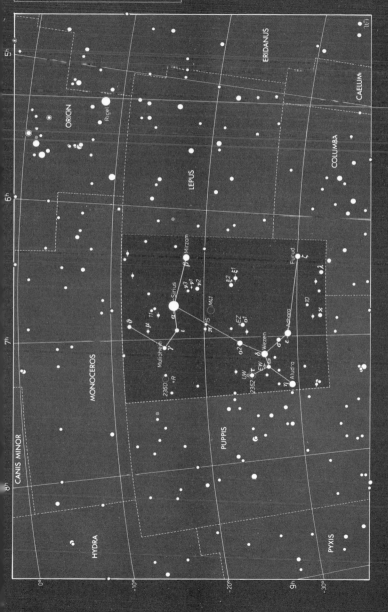

CANIS MINOR   The Lesser Dog

The second of the two dogs of Orion, the other being Canis Major. Apart from its leading star Procyon, one of the brightest stars in the sky, there are few objects of importance in Canis Minor. Procyon forms a prominent equilateral triangle with the bright stars Sirius (in Canis Major) and Betelgeuse (in Orion).

α (alpha) Canis Minoris (Procyon, from the Greek meaning 'before the dog', referring to its rising before Canis Major), mag. 0.38, is a yellow-white star 11.3 l.y. away, and therefore among the nearest stars to the Sun. Like Sirius, Procyon has a white dwarf companion, but this star, of mag. 11, is even more difficult to see than the companion of Sirius, requiring the use of large professional telescopes. Procyon's companion orbits it every 41 years.

β (beta) CMi (Gomeisa), mag. 2.9, is a blue star 140 l.y. away.

White Dwarfs

By a remarkable coincidence both Sirius, the brightest star in Canis Major, and Procyon, the brightest star in Canis Minor, are accompanied by tiny, faint stars known as white dwarfs. The existence of companion stars to Sirius and Procyon was predicted in 1844 by the German astronomer Friedrich Wilhelm Bessel, who detected a wobble in the proper motions (p. 12) of Sirius and Procyon; Bessel realized that this wobble was most probably caused by the presence of unseen companions orbiting around the visible stars. The companion of Sirius, called Sirius B, was first seen in 1862 by the American astronomer Alvan G. Clark using a 47 cm refractor, and the companion to Procyon (Procyon B) was first seen in 1896 by John M. Schaeberle with the 91 cm refractor at Lick Observatory. But not until 1915 was the truly extraordinary nature of these stars realized. Observations showed that Sirius B was very hot, very small, and very dense. In fact Sirius B has the mass of the Sun packed into a sphere only 2 per cent the Sun's diameter. The resulting density of Sirius B is over $100,000\times$ that of water. A white dwarf is a star at the end of its life; it is the shrunken remnant of a once-proud star like the Sun whose central nuclear fires have gone out. The cause of the immense densities of white dwarfs is the inexorable pull of gravity, which squeezes the electrons of the dying star as closely together as is physically possible.

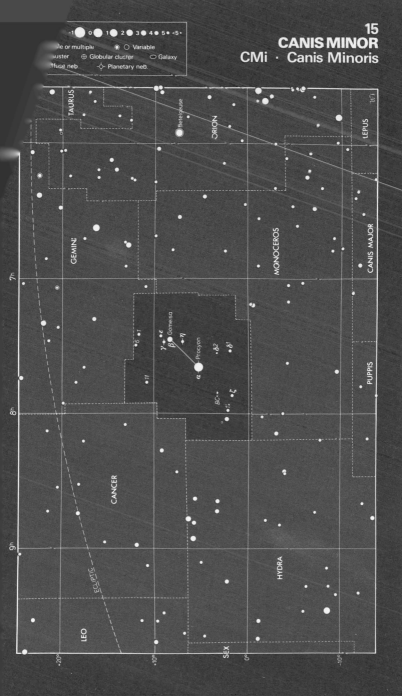

CAPRICORNUS The Sea Goat

Capricornus is depicted as a Goat with a fish tail. Amphibious crea[...]
feature prominently in ancient legends, and the origin of Caprico[...]
certainly dates back to ancient times. About 2500 years ago the [...]
used to reach its farthest point south of the equator in Capricornus
the winter solstice, December 22. The latitude on Earth, 23.5° sout[...]
at which the Sun appears overhead at noon on the winter solstice
therefore became known as the Tropic of Capricorn. Because of precession, the winter solstice has now moved into the neighbouring constellation of Sagittarius, but the Tropic of Capricorn retains its name.

α (alpha) Capricorni (Giedi or Algedi, meaning goat or ibex) is a multiple star,
consisting of two unrelated yellow and orange stars 1600 and 120 l.y. away, of
mags. 4.2 and 3.6 respectively, seen separately with the naked eye or binoculars.
But telescopes reveal that each star is itself double. α^1 (alpha¹), the fainter of
the pair, has a wide 9th mag. companion visible in small telescopes which is
unrelated. α^2 (alpha²) is a genuine binary, with an 11th mag. companion. Telescopes of at least 100 mm aperture show that this faint companion is itself composed of two close 11th mag. stars. α (alpha) Capricorni is therefore a fascinating
hybrid system.

β (beta) Cap (Dabih, lucky one of the slaughterers), mag. 3.1, is a golden yellow
star 250 l.y. away. It has a wide blue companion of mag. 6 visible in binoculars or
small telescopes.

γ (gamma) Cap (Nashira, the fortunate one), mag. 3.7, is a white star 100 l.y. away.

δ (delta) Cap (Deneb Algiedi, goat's tail), mag. 2.9, is the brightest star in the
constellation. It is an eclipsing binary, varying by a barely perceptible 0.2 of a
mag. every 24½ hrs. It lies 49 l.y. away.

π (pi) Cap, 470 l.y. away, is a blue–white star of mag. 5.3 with a close mag. 8.5
companion visible with small telescopes.

M 30 (NGC 7099) is an 8th mag. globular cluster 40,000 l.y. away visible in small
telescopes, notably condensed at the centre, with chains of stars leading outwards
from it.

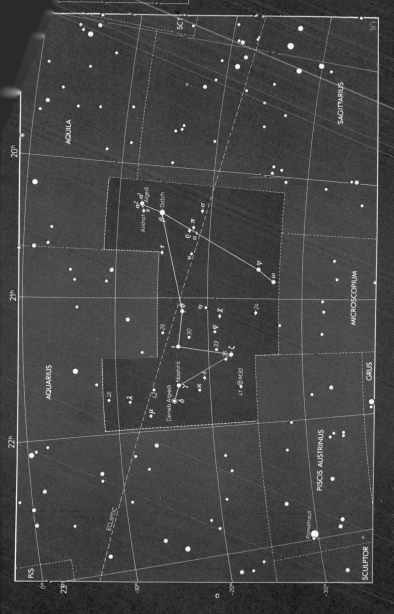

CARINA   The Keel

This constellation was originally part of the extensive Argo Na
Ship of the Argonauts, until subdivided in the 1750s by the F
celestial cartographer Nicolas Louis de Lacaille. As a part of
Navis, Carina's origin goes back to Greek times, and is associated
the legend of Jason and the Argonauts who sought the Golden Flee
Carina lies in the Milky Way, providing rich star fields and clusters f
binoculars. The stars ι (iota) and ε (epsilon) Carinae, together witt
κ (kappa) and δ (delta) Velorum, form the False Cross, sometimes confused with the real Southern Cross.

α (alpha) Carinae (Canopus), mag. −0.72, the 2nd-brightest star in the sky, is a
yellow-white supergiant 1200 l.y. away. It is named after the pilot of the fleet
of the Greek King Menelaos, and appropriately enough this star is now used as a
guide for navigating spacecraft.

β (beta) Car (Miaplacidus), mag. 1.7, is a blue-white star 85 l.y. away.

ε (epsilon) Car, mag. 1.9, is a yellow giant star 200 l.y. away.

η (eta) Car, 9000 l.y. away, is a peculiar nova-like variable star embedded in a
nebula called NGC 3372 (see separate entry below). η (eta) Carinae in the past has
fluctuated erratically in brightness, reaching a maximum of mag. −1 in 1843,
but has now settled to around 6th or 7th mag. The star is surrounded by a shell
of gas thrown off in the 1843 outburst; further variations can come at any time,
and the star is thought to be a candidate for a future supernova.

θ (theta) Car, mag. 2.8, is a blue-white star 750 l.y. away at the centre of the cluster
IC 2602 (see below).

ι (iota) Car, mag. 2.3, is a yellow-white supergiant 820 l.y. away.

υ (upsilon) Car, 320 l.y. away, is a double star consisting of white components of
mag. 3.1 and 6.0, divisible in small telescopes.

R Car is a red giant star of uncertain distance which varies between mags. 4 and
10 every 309 days.

S Car is a red giant variable, similar to R Car, varying from 5th to 10th mag.
every 150 days.

IC 2602 is a large and brilliant cluster of 30 stars similar to the Pleiades, about
700 l.y. away. The brightest members of the cluster are visible to the naked eye,
notably θ (theta) Carinae (see above), and the whole cluster covers an area twice
that of the full Moon.

NGC 2516 is a bright cluster of 100 stars, 4300 l.y. away, visible to the naked eye,
covering 1° of sky. Small telescopes show a central 5th mag. red giant. NGC 2516
contains three double stars of 8th and 9th mag. visible in small telescopes.

NGC 3114 is a naked eye cluster of about 100 stars, approx. 1000 l.y. away, a
poorer version of NGC 2516. ▶

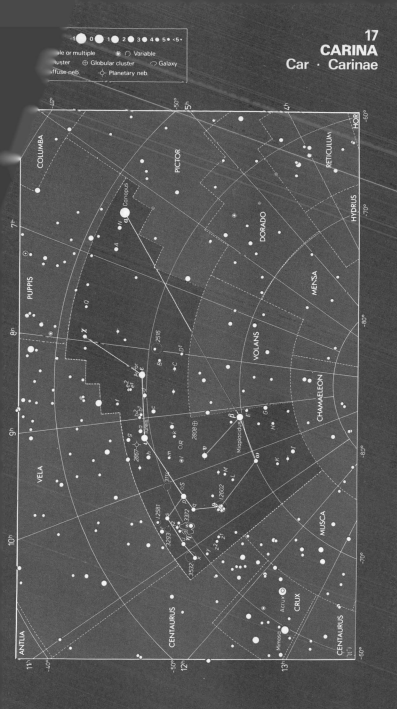

CASSIOPEIA

In legend, Cassiopeia was the beautiful but boastful Queen of Eth▮ wife of King Cepheus and mother of Andromeda. In the sky s depicted sitting in a chair. The constellation is easily spotted by the tinctive W-shape of its five brightest stars, which form part of the ch▮ Cassiopeia lies on the opposite side of the Pole Star from Ursa Major, a rich part of the Milky Way. Near the star κ (kappa) Cassiopeia▮ occurred the famous supernova outburst of 1572, observed by Tycho Brahe. The remains of this supernova now form a radio source, about 20,000 light years away. The remains of another supernova, which erupted around 1660 but which went unseen at the time, form the strongest radio source in the sky, Cassiopeia A; it lies near the star cluster M 52. Cassiopeia A is 10,000 l.y. away.

α (alpha) Cassiopeiae (Schedar, the breast), mag. 2.2, is a yellow giant star 120 l.y. away. It has a wide 9th mag. companion.

β (beta) Cas (Caph), mag. 2.3, is a white star 42 l.y. away.

γ (gamma) Cas, 780 l.y. away, is a remarkable blue giant variable of the type known as a *shell star*. It throws off rings of gas at irregular intervals, apparently because its high-speed rotation makes it unstable, causing it to vary unpredictably between mags. 3.0 and 1.6. Usually it hovers around mag. 2.5.

δ (delta) Cas (Ruchbah, the knee), mag. 2.7, is a blue-white star 62 l.y. away.

ε (epsilon) Cas, mag. 3.4, is a blue giant star 520 l.y. away.

η (eta) Cas, 19 l.y. away, is a beautiful double star with yellow and red components of mags. 3.6 and 7.5, visible in small telescopes.

ι (iota) Cas, 180 l.y. away, is a mag. 4.5 white star with a wide mag. 8 companion visible through a 60 mm telescope. With apertures of at least 100 mm and high magnification, the brighter star is seen to have a closer mag. 7 yellow companion, making this an impressive triple.

σ (sigma) Cas, 1300 l.y. away, is a close pair of mags. 5.1 and 7.2 appearing green ▶

◀ NGC 3372 is a diffuse nebula easily visible to the naked eye as a brilliant patch of the Milky Way, surrounding the erratic variable star η (eta) Carinae. The NGC 3372 Nebula covers an area greater than that of the Orion Nebula. It shines from the light of brilliant young stars that have been born within it. Binocular users will find jewelled star clusters and swirls of glowing gas alternating with dark lanes in this nebula. Most famous is a dark notch, called the Keyhole because of its distinctive shape, silhouetted against the nebula's brilliant central portion near η (eta) Carinae itself. The whole NGC 3372 Nebula lies about 9000 l.y. from us, the same distance as η (eta) Carinae.

NGC 3532, 1700 l.y. away, is a bright cluster of 150 stars covering 1° of sky, visible to the naked eye and glorious in binoculars.

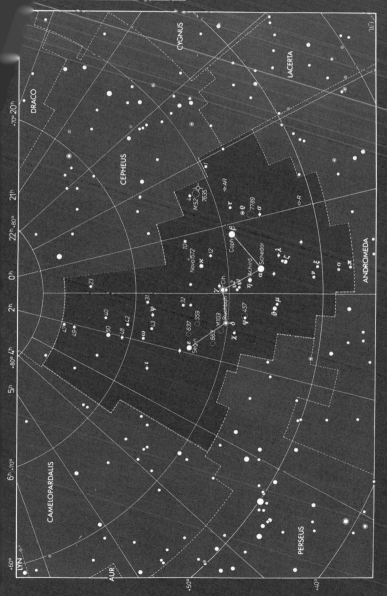

CENTAURUS The Centaur

A large and rich constellation representing a centaur in Greek m_
ology, reputedly the scholarly centaur Chiron who was the tutor
many Greek heroes, and who according to legend invented the m_
constellations. He was raised to the sky after being accidentally struc
by a poisoned arrow from Hercules. (This legend is sometimes also tol_
for Sagittarius, another celestial centaur.) Centaurus is of particular
interest because it contains the closest star to the Sun, α (alpha) Centauri, which is actually a group of three stars linked by gravity. A line
from α (alpha) through β (beta) Centauri points to Crux, the Southern
Cross. One of the strongest radio sources in the sky, Centaurus A, lies
in the constellation, associated with the galaxy NGC 5128. Centaurus
lies in a prominent part of the Milky Way.

α (alpha) Centauri (Rigil Kentaurus, foot of the centaur) lies 4.3 l.y. away. To
the naked eye it shines as a star of mag. −0.27, 3rd-brightest in the sky, but the
smallest of telescopes reveals that it consists of two individual yellow stars of
mags. 0 and 1.4. The brighter of these is very similar in nature to the Sun. They
orbit each other every 80 years. Also associated with α (alpha) Centauri is an 11th
mag. red dwarf called Proxima Centauri, lying 2° away and therefore not even in
the same telescopic field of view. This star is estimated to take as much as a
million years to orbit its two brilliant companions. At present, Proxima Centauri
is about 0.1 l.y. closer to us than the two other members of α (alpha) Centauri. ▶

◀ and blue, in striking contrast to the warmer hues of η (eta) Cas. Apertures of
75 mm and high power will split the pair, but 150 mm is needed to show the
colours.

ψ (psi) Cas, 240 l.y. away, is a mag. 5 yellow giant with a wide 9th mag. companion visible in small telescopes. High powers reveal that this companion is
itself a close binary.

M 52 (NGC 7654) is a cluster of about 120 stars, 3800 l.y. away, visible as a misty
patch in binoculars. It is somewhat kidney-shaped, with a prominent yellow-
orange star embedded at one edge, like a poor version of the celebrated Wild Duck
Cluster (M 11 in Scutum). M 52 can be resolved into individual stars with apertures of 75 mm.

M 103 (NGC 581) is a scattering of about 60 stars with a central diamond shape,
3800 l.y. away.

NGC 457 is a loose cluster of faint stars, seemingly arranged in chains. The mag.
5.0 yellow supergiant star φ (phi) Cas at one side acts as a certain guide to the
cluster.

NGC 663 is a binocular cluster of about 80 stars, 2600 l.y. away.

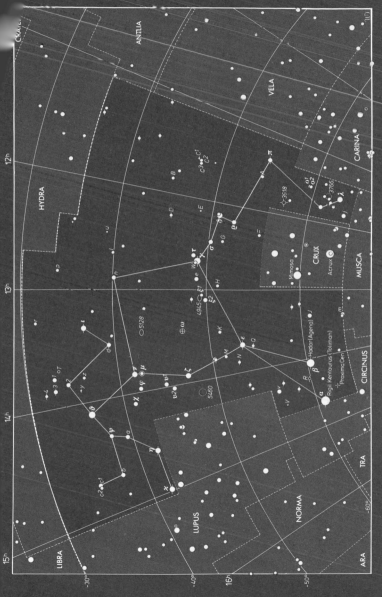

ω (omega) Centauri, the finest of all globular star clusters, appears noticeably elliptical in this photograph. *Royal Observatory, Edinburgh.*

◀ Proxima Centauri is a flare star, suddenly increasing in brightness by as much as one mag. for several minutes. (See finder chart for Proxima Centauri.)

β (beta) Cen (Hadar or Agena), mag. 0.6, is a blue giant 460 l.y. away.

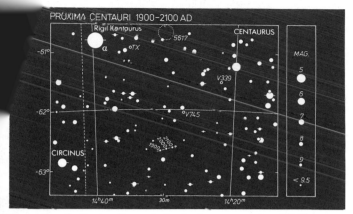

Finder chart for Proxima Centauri, showing its proper motion. *Wil Tirion.*

γ (gamma) Cen, 110 l.y. away, is a close double with blue-white components of mags. 3.1 and 3.2 orbiting every 85 years. Together they shine as a star of mag. 2.2. At their widest in 1973 they were visible in apertures of 100 mm, but when they are at their closest around the year 2015 amateur telescopes will be unable to split them.

3 Cen, also known as k Cen, 300 l.y. away, is a striking pair of blue-white stars of mags. 4.7 and 6.2, divisible in small telescopes.

ω (omega) Cen (NGC 5139) is the largest and brightest globular cluster in the sky – so prominent that it was labelled as a star on early charts. It appears of mag. 3.7 to the naked eye, and covers an area over two-thirds that of the full Moon. Small telescopes or even binoculars begin to resolve its outer regions into stars, and it is a showpiece for all apertures. Its brilliance and large apparent size are due in part to its relative closeness of 16,000 l.y., which makes it among the nearest globular clusters to us.

NGC 3766 is a naked eye cluster of about 60 stars, 1700 l.y. away.

NGC 3918 is a small, 8th mag. planetary nebula 14,500 l.y. away, discovered by John Herschel and called by him the Blue Planetary. It is similar in appearance to the planet Uranus but considerably larger.

NGC 5128 is a peculiar 7th mag. galaxy known to radio astronomers as Centaurus A. In long-exposure photographs it appears as a giant elliptical galaxy with an encircling band of dust, but on radio maps it is seen flanked by lobes of radio emission as though having ejected matter from a series of explosions. Under good skies it is visible in binoculars, but at least 100 mm aperture is necessary to trace its outline and the dark bifurcating lane of dust. NGC 5128 is about 15 million l.y. away.

NGC 5460 is a large cluster of about 25 stars visible in binoculars or small telescopes. It lies 2700 l.y. away.

CEPHEUS

An ancient constellation representing King Cepheus of Ethi...
husband of Cassiopeia and father of Andromeda, themselves represen...
by nearby constellations. Cepheus is situated on the edge of the Mi...
Way, and although it contains no spectacular star clusters or nebul...
it is replete with double and variable stars, including the celebrate...
δ (delta) Cephei, prototype of the Cepheid variables, used as 'standar...
candles' for distance-finding in space. This star's fluctuations in light
output were discovered in 1784 by the English amateur astronomer John
Goodricke, a deaf-mute who died in 1786 at the age of 21.

α (alpha) Cephei (Alderamin, right shoulder), mag. 2.4, is a white star 46 l.y. away.

β (beta) Cep (Alfirk, meaning herd), 750 l.y. away, is both a double and a variable
star. A small telescope shows that this blue giant star, of mag. 3.2, has an 8th mag.
companion. β (beta) Cephei is the prototype of a class of pulsating variable stars
(also known as β (beta) Canis Majoris stars) with periods of a few hours and tiny
light fluctuations. Every 4½ hrs. or so, β (beta) Cephei varies by less than 0.1 of a
mag., an amount indistinguishable to the naked eye but which can be detected
by sensitive instruments.

γ (gamma) Cep (Er Rai, shepherd), mag. 3.2, is a yellow star 52 l.y. away.

δ (delta) Cep, 1300 l.y. away, is a famous pulsating variable star, prototype of the
classic Cepheid variables. This yellow giant varies between mags. 3.6 and 4.3
every 5 days 9 hrs., changing in size from 35–32 × the Sun's diameter as it does
so. Less well known is that δ (delta) Cephei is also an attractive double star for
binoculars or the smallest telescopes, with a wide, bluish 6th mag. companion.

μ (mu) Cep, 1600 l.y. away, is a famous red-coloured star, called the Garnet Star
by Sir William Herschel because of its striking tint, which is notable in binoculars.
μ (mu) Cephei is a red supergiant, the prototype of a class of variable stars known
as semiregular variables. It varies between mags. 3.6 and 5.1 with no set period.

ξ (xi) Cep, 120 l.y. away, is a double star of mags. 4.6 and 6.6 visible in small
telescopes. The components are blue-white and yellow.

o (omicron) Cep, 260 l.y. away, is a yellow giant of mag. 5.0 with a close mag. 7.3
companion for telescopes of 60 mm aperture and above.

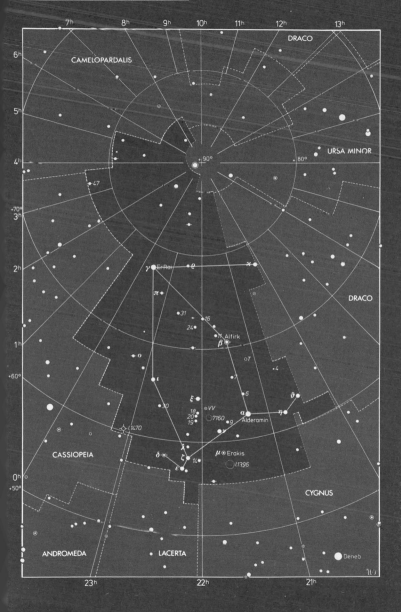

Magnitudes: -1 ● 0 ● 1 ● 2 ● 3 ● 4 ● 5 ● <5 ·
● ● Double or multiple   ◉ ○ Variable
○ Open cluster   ⊕ Globular cluster   ○ Galaxy
□ ⌒ Diffuse neb.   ⋄ Planetary neb.

DRACO

CAMELOPARDALIS

URSA MINOR

·90°   ·80°

·47

DRACO

γ ErRai   ℓ   ×

π

31
24   16

η Alfirk
β

τ

4

7

6   ϑ

ξ   η

ι   α   VV   Alderamin   η

30   18   7160
20   9
19

λ   μ Erakis

δ   14   LL396
ζ
ε

CASSIOPEIA

CYGNUS

LL470

ANDROMEDA   LACERTA   Deneb

CETUS   The Whale

An ancient constellation depicting the sea monster that threatened to devour Andromeda before she was rescued by Perseus. Cetus is found in the sky basking on the shores of Eridanus, the River. The constellation is large but faint, containing several stars of particular interest, notably *o* (omicron) Ceti and *τ* (tau) Ceti. One faint but famous star is UV Ceti, which consists of a pair of red dwarfs 8.9 l.y. away, one of which is the prototype of a class of erratic variables known as *flare stars*; these are red dwarfs which undergo sudden increases in light output lasting a few minutes. The outbursts of the flare star component of UV Ceti take it from its normal level of 13th mag. to as bright as 7th mag. ▶

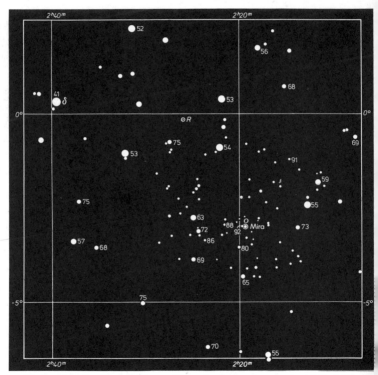

Finder chart for the variable star Mira, also known as *o* (omicron) Ceti. The numbers by the surrounding stars are not Flamsteed numbers but their magnitudes, with the decimal point omitted. This convention is followed on such charts to prevent confusion between decimal points and faint stars. The magnitude of Mira can be estimated by comparison with the surrounding stars. *Wil Tirion.*

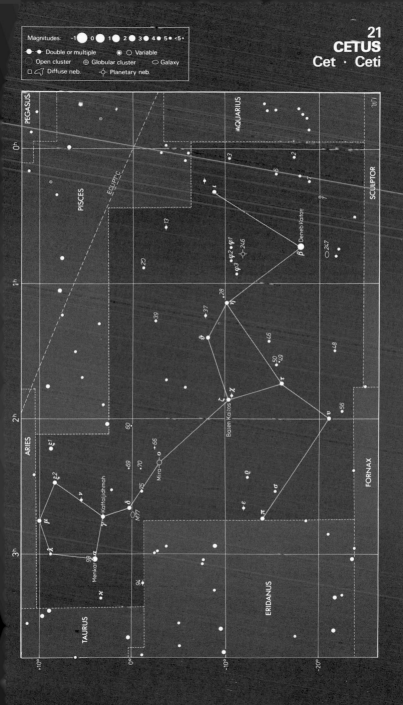

Magnitudes: -1 0 1 2 3 4 5 <5
● ● Double or multiple   ◉ ○ Variable
○ Open cluster   ⊕ Globular cluster   ○ Galaxy
□ ◁ Diffuse neb.   ◇ Planetary neb.

PEGASUS

AQUARIUS

PISCES

ECLIPTIC

SCULPTOR

ι

13

β Deneb Kaitos

φ3 φ2 φ1

246

247

20

28

η

39

37

θ

46

50
49

τ

48

ζ

χ

Baten Kaitos

ν

56

60

ξ1

ξ2

ν

Kaffaljidhmah

69
70

75

δ

66
o

Mira ○

θ

55

ε

σ

π

μ

γ

M77

ARIES

λ

α

Menkar

91

94

χ

FORNAX

ERIDANUS

TAURUS

CHAMAELEON

A faint and undistinguished constellation, representing a chameleon. It was introduced into the southern skies by Johann Bayer in 1603.

α (alpha) Chamaeleontis, mag. 4.1, is a white star 78 l.y. away.

β (beta) Cha, mag. 4.3, is a blue star 360 l.y. away.

γ (gamma) Cha, mag. 4.1, is a red giant 250 l.y. away.

δ (delta) Cha consists of a wide pair of unrelated stars, clearly seen in binoculars. $δ^1$ (delta^1) Cha, mag. 5.5, is an orange star 360 l.y. away. $δ^2$ (delta^2) Cha, mag. 4.5, is a blue star 550 l.y. away.

ε (epsilon) Cha, 290 l.y. away, consists of a very close pair of stars of mag. 5.5 and 6.3, needing apertures of 150 mm or above with high magnification to split.

NGC 3195 is a planetary nebula of similar apparent size to the planet Jupiter.

◄ α (alpha) Ceti (Menkar, meaning nose), mag. 2.5, is a red giant star 130 l.y. away. Binoculars show a wide mag. 5.6 blue companion which is unrelated, lying over 6× farther away.

β (beta) Cet (Deneb Kaitos, tail of the whale), mag. 2.0, the brightest star of the constellation, is a yellow giant 68 l.y. away.

γ (gamma) Cet (Kaffaljidhmah), 75 l.y. away, is a close double star needing telescopes of at least 60 mm aperture and high power to split. The stars are of mags. 3.7 and 6.4, the colours yellow and bluish.

o (omicron) Cet (Mira, the wonderful), 820 l.y. away, is the prototype of a famous class of red giant long-period variable stars. Mira itself varies between about 3rd–9th mag. in an average of 331 days, changing in diameter from 300–400× the size of the Sun. Mira's light variations were first noted in 1596 by the Dutch astronomer David Fabricius, making it the first variable star to be discovered. (See finder chart for Mira.)

τ (tau) Cet, mag. 3.5, a yellow dwarf, is one of the nearest stars to the Sun, lying 11.7 l.y. away. Its main claim to fame is that, of all the nearby single stars, it is the one most like the Sun. Some astronomers speculate that it may have planets, although there is as yet no evidence of their existence. Radio astronomers have listened for possible alien radio messages coming from τ (tau) Ceti, but without success.

M 77 (NGC 1068) is a small, softly glowing 9th mag. face-on spiral galaxy located near δ (delta) Ceti, requiring telescopes of 150 mm aperture or above to be well seen. It lies approx. 50 million l.y. away, and is perhaps the most remote of the objects on Messier's list. M 77 is classified as a Seyfert galaxy, a type of spiral with a brilliant nucleus, and it is a radio source.

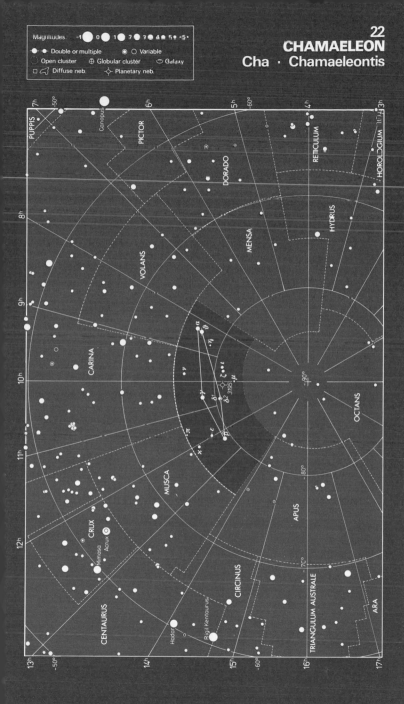

Magnitudes: -1 0 1 2 3 4 5 <5

●—● Double or multiple   ◉ ○ Variable
◐ Open cluster   ⊕ Globular cluster   ⬭ Galaxy
▢ ◁ Diffuse neb.   ◇ Planetary neb.

CIRCINUS   The Compasses

Another of the small and obscure southern constellations introduced by the French astronomer Nicolas Louis de Lacaille. It represents a pair of compasses, appropriately placed in the sky next to Norma, the Level. It is overshadowed by the brilliant neighbouring Centaurus.

α (alpha) Circini, 46 l.y. away, is a white star of mag. 3.2 with a wide mag. 8.8 companion visible in small telescopes.

γ (gamma) Cir, 270 l.y. away, consists of a very close pair of blue and yellow 5th mag. stars, visible separately in telescopes of 150 mm aperture under high magnification.

In the neighbouring constellation of Norma lies the satisfyingly symmetrical 8th mag. planetary nebula known as Shapley 1 (*Sp 1*), surrounding a 14th mag. central star. *Anglo-Australian Telescope Board.*

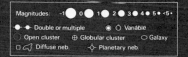

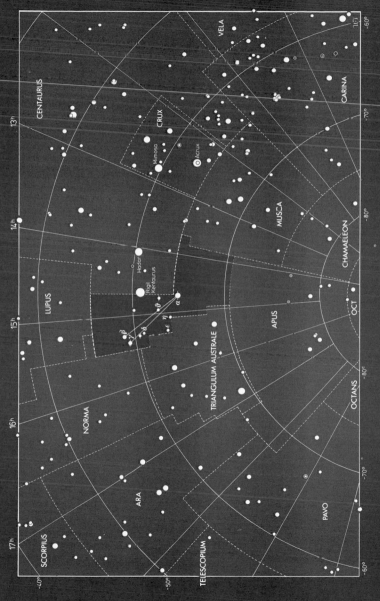

Magnitudes: -1 0 1 2 3 4 5 <5

- Double or multiple   Variable
- Open cluster   Globular cluster   Galaxy
- Diffuse neb.   Planetary neb.

VELA

CENTAURUS

CARINA

CRUX

Mimosa

Acrux

MUSCA

CHAMAELEON

Hadar

Rigil
Kentaurus

θ
α

LUPUS

η

β
δ
ε

APUS

OCT

TRIANGULUM AUSTRALE

OCTANS

NORMA

ARA

PAVO

SCORPIUS

TELESCOPIUM

COLUMBA   The Dove

A constellation adjoining Puppis, representing the dove that followed Noah's Ark, or the dove that the Argonauts sent ahead to help them pass safely between the Symplegades, the Clashing Rocks, at the mouth of the Black Sea. Columba has been an accepted constellation since 1679, when it was included on a list by the Frenchman Augustin Royer, but before him it had been introduced on star charts by others, notably Johann Bayer. It contains no objects of interest for amateur telescopes.

α (alpha) Columbae, mag. 2.6, is a blue-white star 120 l.y. away.

β (beta) Col, mag. 3.1, is a yellow giant 140 l.y. away.

In Coma Berenices lies the classically elegant NGC 4565, a spiral galaxy seen exactly edge-on (see page 118). Note the central bulge of stars and the dark lanes of dust in the galaxy's plane. *Hale Observatories photograph.*

Magnitudes: -1 0 1 2 3 4 5 <5
● ● Double or multiple ◉ ○ Variable
○ Open cluster ⊕ Globular cluster ◯ Galaxy
□ ▱ Diffuse neb. ◇ Planetary neb.

24
COLUMBA
Col · Columbae

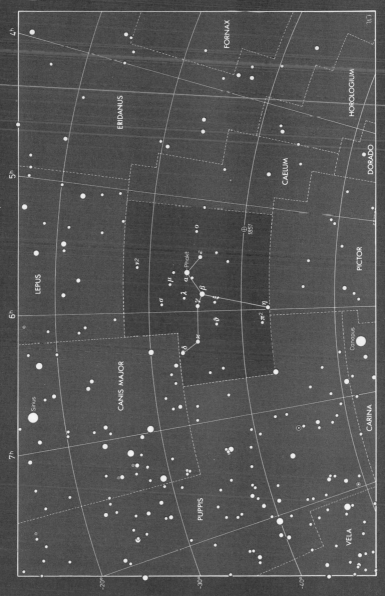

COMA BERENICES   Berenice's Hair

This faint constellation represents the flowing locks of Queen Berenice of Egypt, who cut off her hair in gratitude to the gods for the safe return from battle of her husband Ptolemy Euergetes. Although this legend dates from Greek times, the constellation did not become firmly established until 1602 when it was included in a catalogue by the Danish astronomer Tycho Brahe. The main part of the Queen's severed tresses is represented by a scattered group of about 30 stars known as the Coma Star Cluster, covering several degrees of sky and best seen in binoculars. The brightest members of the star cluster, of 5th and 6th mag., form a noticeable V-shape. The stars of the cluster, about 250 light years distant, like near γ (gamma) Comae, which itself is apparently not a member of the cluster. Coma Berenices also contains another type of cluster – a cluster of galaxies. The Coma Cluster of galaxies (not to be confused with the Coma Star Cluster) lies about 400 million light years away, so its members are too faint for all but the largest amateur telescopes. But the constellation also contains some brighter galaxies, members of the nearer Virgo Cluster, the brightest of which are visible in amateur telescopes. The north pole of the galaxy lies in Coma Berenices.

α (alpha) Comae Berenicis (Diadem), mag. 4.3, is a white star 59 l.y. distant.

β (beta) Com, mag. 4.3, is a yellow star 27 l.y. away.

γ (gamma) Com, mag. 4.4, is an orange star 300 l.y. away.

24 Com, 360 l.y. away, is a beautiful coloured double star for small telescopes, consisting of an orange giant of mag. 5.2 and a mag. 6.7 blue-white companion.

M 53 (NGC 5024) is an 8th mag. globular cluster 65,000 l.y. away, visible in small telescopes as a rounded, hazy patch.

M 64 (NGC 4826) is a famous spiral galaxy, called the Black Eye Galaxy because of a dark cloud of dust bordering its nucleus. This dark dust lane shows up well in telescopes above 150 mm aperture; observers with smaller instruments must content themselves simply with locating this 9th mag. galaxy 20 million l.y. away.

M 85 (NGC 4382) is a 9th mag. elliptical galaxy 65 million l.y. away. Small telescopes show a bright, star-like centre.

M 88 (NGC 4501) is a 10th mag. spiral galaxy seen nearly edge-on, so that it appears elliptical. It is 65 million l.y. away.

M 99 (NGC 4254) is a 10th mag. spiral 65 million l.y. away, seen face-on so that it appears almost circular.

M 100 (NGC 4321), 65 million l.y. away, is a 10th mag. spiral seen face-on, similar to M 99 but larger.

NGC 4565 is a 10th mag. galaxy seen edge-on. It is the most famous of the edge-on ▶

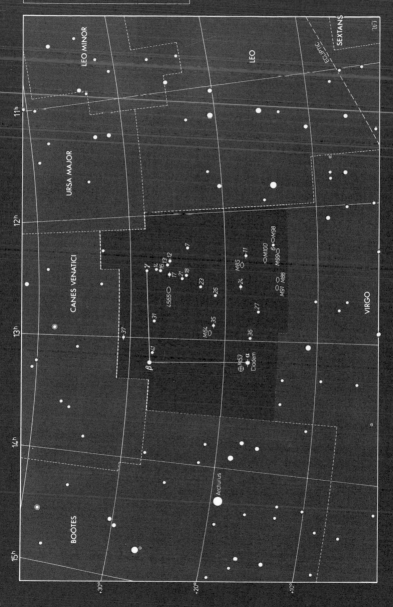

CORONA AUSTRALIS   The Southern Crown

The southern counterpart to the Northern Crown (Corona Borealis). Corona Australis has been known since the time of the Greek astronomer Ptolemy in the 2nd Century AD. In one legend it is said to represent the crown worn by the centaur Sagittarius, which it lies next to. Although faint, it is a noticeable figure, situated on the edge of the Milky Way.

α (alpha) Coronae Australis, mag. 4.1, is a white star 100 l.y. away.

β (beta) CrA, mag. 4.1, is a yellow giant 110 l.y. away.

γ (gamma) CrA, 39 l.y. away, consists of a pair of identical 5th mag. yellow stars orbiting every 120 years, forming a tight double for small telescopes. As seen from Earth, the two stars are closest together around 1990, when at least 100 mm aperture will be needed to split them.

κ (kappa) CrA, 460 l.y. away, is a pair of 6th mag. blue-white stars easily visible in small telescopes.

NGC 6541 is a 6th mag. globular cluster 14,000 l.y. away, visible in binoculars or small telescopes.

◀ spirals, and is pictured on page 116. Instruments of 200 mm aperture and above show it as a cigar-shaped body split by a dark band (actually a dust lane), and a noticeable central bulge with a star-like core. Its distance is about 20 million l.y.

Many of the brightest galaxies in Coma Berenices are members of the Virgo Cluster, the centre of which lies just south of the Coma-Virgo border. At the cluster's heart lies the massive elliptical galaxy M 87, which photographs show as shooting out a long jet of hot gas. See page 252. *Lick Observatory.*

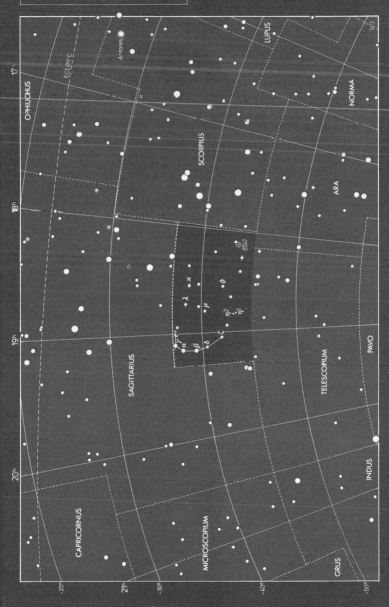

CORONA BOREALIS The Northern Crown

An ancient constellation representing the jewelled crown given as a wedding present by Bacchus to Ariadne, and cast by him into the sky when she died. In another legend it is the golden crown given to Ariadne by Theseus after their adventure into the labyrinth to kill the Minotaur. It has also been seen as a laurel wreath. It consists of an arc of seven stars, all but one being of 4th magnitude; the exception is 2nd magnitude Gemma, which is set in the crown like a central gem, as its name implies. Corona Borealis contains a famous cluster of about 400 galaxies more than 1000 million light years away. Being so very distant, the galaxies are no brighter than 16th magnitude and thus are far beyond the reach of amateur telescopes.

α (alpha) Coronae Borealis (Gemma or Alphecca), mag. 2.2, is a blue-white star 78 l.y. distant. It is an eclipsing binary of the Algol type, but its variation every 17.4 days is only 0.1 of a mag., too slight to be noticeable to the naked eye.

ζ (zeta) CrB, 420 l.y. away, is a pair of blue stars, of mags. 5.0 and 6.0, visible in small telescopes.

ν (nu) CrB is a wide binocular pair of orange giant stars, each of mag. 5.3 but unrelated – one is 490 l.y. distant and the other is 420 l.y. away.

σ (sigma) CrB, 68 l.y. away, is a pair of yellow stars for small telescopes, of mags. 5.7 and 6.7.

R CrB is a remarkable star lying within the arc of the Crown, halfway between the stars α (alpha) and ι (iota). It usually appears about 6th mag., but occasionally and unpredictably drops in a matter of weeks to as faint as mag. 14, from where it may take many months to regain its former brightness. Recent catastrophic declines in the star's brilliance occurred in 1962, 1972 and 1977, and another could occur at any time. These sudden dips in the light of R Coronae Borealis are believed to be due to the accumulation of graphite particles in its atmosphere. R Coronae Borealis is thought to be a yellow supergiant; its distance is uncertain.

T CrB is another spectacular star, known as the Blaze Star, which performs in almost the opposite way to R Coronae Borealis. It is a recurrent nova, usually slumbering at around mag. 11 but which can suddenly and unpredictably brighten to mag. 2. Its last recorded outburst occurred in 1946, and the previous one was 80 years before that. It is not known when it may erupt again.

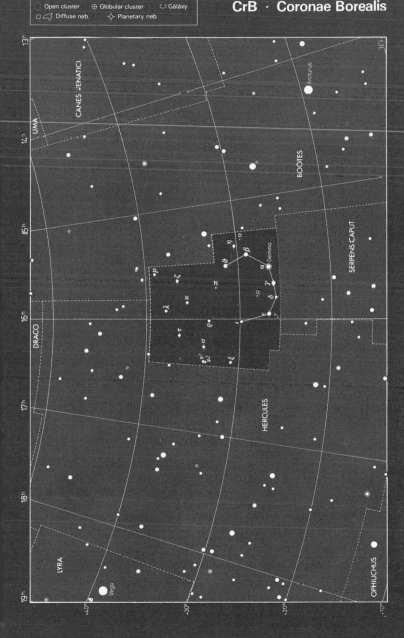

CORVUS   The Crow

The origin of this faint constellation dates back at least to the ancient
Greeks, who called it the Raven. In legend, Corvus is associated with
the neighbouring constellations of Crater, the Cup, and Hydra, the
Water Snake. The crow is said to have been sent by Apollo to gather
water in a cup; but instead it dallied to eat figs. When the crow returned
to Apollo it carried the water snake in its claws, claiming this creature
to have been the cause of its delay. Apollo, realizing the lie, banished
the trio to the sky, where the Crow and the Cup lie on the back of Hydra.
For its misdeed the crow was condemned to suffer from eternal thirst,
which is why crows croak so harshly. In another legend, a snow-white
crow brought Apollo the bad news that his lover Coronis had been un-
faithful to him. In his anger, Apollo turned the crow black. Apollo and
crows are closely linked in legend, for during the war waged by the
giants with the gods, Apollo turned himself into a crow.

α (alpha) Corvi (Al Chiba), mag. 4.0, is a white star 68 l.y. away.

β (beta) Crv, mag. 2.7, is a yellow giant star 290 l.y. away.

γ (gamma) Crv, mag. 2.6, is a blue-white star 190 l.y. away.

δ (delta) Crv, 120 l.y. away, is a wide double star for small telescopes. The brightest
component, visible to the naked eye, is a mag. 3.0 white star, which is accompanied
by a mag. 8.4 star often described as purplish in colour.

Just over the border of Corvus with Virgo lies the Sombrero Galaxy, M 104,
looking rather like the planet Saturn. It is a spiral galaxy with a large nucleus and
a dense lane of dust. See also page 252. *Anglo-Australian Telescope Board.*

Magnitudes: -1 0 1 2 3 4 5 • <5 •
●–●–● Double or multiple   ◉ Variable
◯ Open cluster   ⊕ Globular cluster   ◯ Galaxy
▢ ⌓ Diffuse neb.   ◇ Planetary neb.

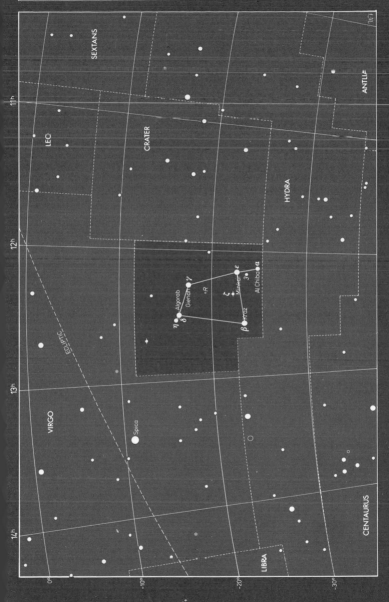

SEXTANS

ANTLIA

11ʰ

LEO

CRATER

HYDRA

12ʰ

ε
γ
Gienah   Minkar
Algorab   •R   ζ   3
Al Chiba α
η
δ   β   Kraz

ECLIPTIC

13ʰ

VIRGO

Spica

14ʰ

CENTAURUS

LIBRA

0°   -10°   -20°   -30°

CRATER   The Cup

An ancient constellation representing the goblet of Apollo, and associated in legend with neighbouring Corvus. It contains no objects of particular interest.

α (alpha) Crateris, mag. 4.1, is a yellow giant star 160 l.y. away.

β (beta) Crt, mag. 4.5, is a blue–white star 230 l.y. away.

γ (gamma) Crt, mag. 4.1, is a white star 78 l.y. away.

δ (delta) Crt, mag. 3.6, the constellation's brightest star, is a yellow giant lying 130 l.y. away.

In neighbouring Corvus, near the border with Crater, lies the peculiar pair of 11th mag. galaxies NGC 4038-9, known as the Antennae because of their long tails. The two galaxies seem to have nearly collided. *Royal Observatory Edinburgh*.

Magnitudes: −1 ● 0 ● 1 ● 2 ● 3 ● 4 ● 5 ● <5 ●
● ● Double or multiple   ◉ ○ Variable
○ Open cluster   ⊕ Globular cluster   ⬭ Galaxy
□ ◁ Diffuse neb.   ◇ Planetary neb.

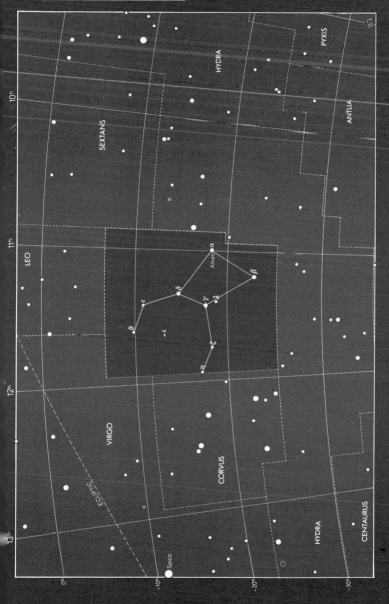

CRUX   The Southern Cross

The smallest constellation in the sky, but one of the most celebrated and distinctive. It was formed from some of the stars of Centaurus by various seamen and astronomers in the 16th Century. Crux lies in a dense and brilliant part of the Milky Way, which makes the famous dark nebula known as the Coalsack seem even more striking in silhouette against the star background. The Coalsack Nebula, which spills over into neighbouring Centaurus and Musca, lies 400 light years away and covers nearly $7° \times 5°$ of sky. Crux, like Centaurus, was visible from the Mediterranean area in ancient times, so that its stars were known to astronomers of Greek civilization; the effects of precession have since carried it below the horizon from such northerly latitudes.

α (alpha) Crucis (Acrux), 360 l.y. away, appears to the naked eye as a star of mag. 0.9. Small telescopes show it to be a double star with blue-white components of mags. 1.4 and 1.9.

β (beta) Cru, mag. 1.3, is a blue-white star 570 l.y. away. It is a variable star of the β (beta) Cephei type, fluctuating by less than 0.1 of a mag. every 6 hrs.

γ (gamma) Cru, mag. 1.6, is a red giant star 88 l.y. away. There is a very wide mag. 6.4 companion, which is unrelated, visible in binoculars.

δ (delta) Cru, mag. 2.8, the faintest of the four main stars in the Cross, is a white star 470 l.y. away.

ε (epsilon) Cru, mag. 3.6, is a yellow star 125 l.y. away.

ι (iota) Cru, 280 l.y. away, is an orange star of mag. 4.7 with a wide mag. 7.8 companion visible in small telescopes.

μ (mu) Cru, 620 l.y. away, is a wide pair of blue-white stars of mags. 4.0 and 5.2, split by the smallest telescopes or even good binoculars.

NGC 4755, the Jewel Box or κ (kappa) Crucis Cluster, is one of the finest star clusters in the sky, visible to the naked eye as a 5th mag. star. Small telescopes show at least 50 stars of various colours, mostly blue and red. John Herschel gave this cluster its popular name when he likened it to a piece of multicoloured jewellery. The distance of the Jewel Box Cluster is uncertain; one measurement places it over 7000 l.y. away.

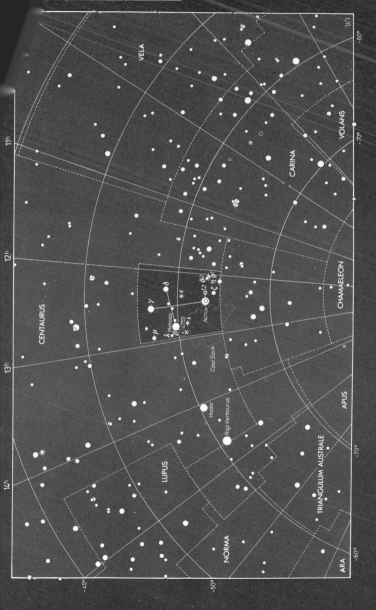

CYGNUS   The Swan

Cygnus represents a swan flying down the Milky Way. Swans in many legends, including that of Zeus who visited Leda, wife o͡ Tyndareus of Sparta, in the guise of a swan; the result of their ͙ was Pollux, one of the heavenly Twins. Another legend iden͙ Cygnus with Orpheus, who was turned into a swan and placed nex͡ his Harp (Lyra) in the sky. But this constellation has also been depict͡ as a hen, notably by the Arabs. The flying Swan's tail is marked b͡ Deneb, its head by Albireo, and its wings by δ (delta) and ε (epsilon͙ Cygni. The constellation is also frequently referred to as the Northern Cross because of its cruciform shape, far larger and more easily recognized than its southern counterpart. Cygnus lies in a particularly rich part of the Milky Way which on dark nights can be seen to be split by a dark lane of dust known as the Cygnus Rift, or the Northern Coalsack.

Deneb, the constellation's brightest star, forms one corner of the famous Summer Triangle completed by Altair and Vega. Some of the most fascinating celestial objects lie in Cygnus. Among them is an X-ray source called Cygnus X-1 which is believed to mark the position of a black hole; it lies near the star η (eta) Cygni. Cygnus A, near γ (gamma) Cygni, is a powerful radio source, believed to be an exploding galaxy far off in space. Long-exposure photographs reveal beautiful swirls of gas, known as the Cygnus Loop, in the region between ε (epsilon) Cygni and the border with Vulpecula. The brightest part of the loop is known as the Veil Nebula. The loop is believed to be the remains of a star that exploded as a supernova about 60,000 years ago.

α (alpha) Cygni (Deneb, meaning tail), mag. 1.3, is a blue-white supergiant star 1800 l.y. away.

β (beta) Cyg (Albireo), 390 l.y. away, is one of the sky's showpiece doubles. It consists of a glorious contrast of yellow and blue-green stars. The brighter star, of mag. 3.1, is a yellow giant, and its companion is mag. 5.1. They can be separated through good binoculars, and are a beautiful sight in any amateur telescope.

γ (gamma) Cyg (Sadr, breast), mag. 2.2, is a yellow supergiant 750 l.y. away.

δ (delta) Cyg, 160 l.y. away, is a blue-white star of mag. 3.0 with a close mag. 6.5 companion, visible in telescopes of 100 mm aperture or above at high magnification.

ε (epsilon) Cyg (Gienah, wing), mag. 2.5, is a yellow giant 82 l.y. away.

μ (mu) Cyg, 55 l.y. away, is a pair of white stars of mags. 4.8 and 6.1 orbiting each other about every 500 years, requiring apertures of 100 mm and above with high magnification to be split. A wide 7th mag. binocular companion is unrelated, but makes this an apparent triple in telescopes.

o^1 (omicron[1]) Cyg is perhaps the most beautiful binocular double in the heavens, consisting of companions of orange and turquoise, mags. 3.8 and 4.8, 520 l.y. and

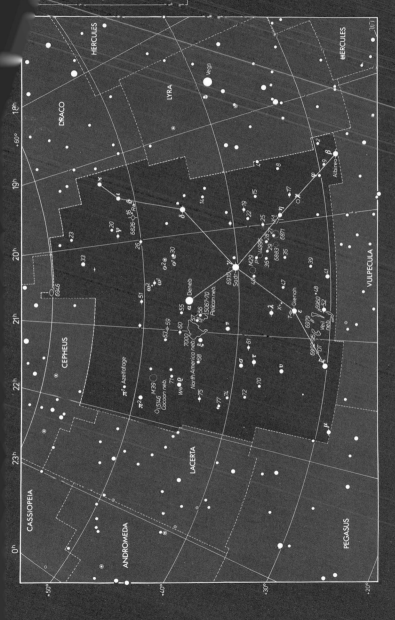

◀ 300 l.y. away, like a wider version of Albireo. Small telescopes or binocula[...]
very still show a closer blue companion of mag. 7 to the brighter (orange,[...]

χ (chi) Cyg, 82 l.y. away, is a red giant long-period variable which varies bet[...]
about 4th or 5th mag. and 13th mag. every 407 days.

ψ (psi) Cyg, 170 l.y. away, is a pair of white stars of mags. 4.9 and 7.4 for smal[...]
moderate apertures.

26 Cyg, 550 l.y. away, is an orange giant of mag. 5.1 with a wide, unrelated 9t[...]
mag. companion for small telescopes.

61 Cyg, 11.1 l.y. away, is a showpiece pair of orange dwarf stars of mags. 5.2 and
6.0, orbiting each other roughly every 700 years. Both can be seen in a small
telescope. In addition to being among the closest stars to Earth, 61 Cygni was the
first star to have its parallax measured, by the German astronomer Friedrich
Wilhelm Bessel in 1838.

P Cyg is a variable blue supergiant which can flare up unpredictably to 3rd mag.;
it normally resides around 5th mag. Evidently the star is so large and luminous
that it is close to being unstable. It lies 4700 l.y. away.

M 39 (NGC 7092) is a large, loose cluster of about 24 fairly bright stars, visible in
binoculars; it lies about 800 l.y. away.

NGC 6826 is an 8th mag. planetary nebula known as the Blinking Planetary
Nebula because it appears to blink on and off. The nebula is visible as a pale blue
disk in 75 mm telescopes, but 150 mm aperture is needed to show it to advantage.
At its centre is a 10th mag. star. Looking alternately at this star and away from it
produces the 'blinking' effect. The nebula lies within 1° of the double star 16
Cygni, consisting of two 6th mag. creamy white stars.

NGC 7000, the North America Nebula, can be seen with binoculars on a clear,
dark night. Despite its large size – up to 2° across – it is difficult to detect because
of its low surface brightness. It is famous from long-exposure photographs which
clearly show its shape, like that of the continent of North America. The nebula
lies about 1500 l.y. away, similar to the distance of the star Deneb, with which it
may be associated.

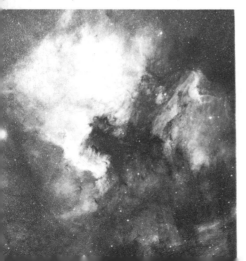

The North America Nebula is a
cloud of glowing gas shaped like
the continent of North America,
its particularly distinctive Gulf of
Mexico being formed by an
obscuring cloud of dark dust. The
bright star at the left is ξ (xi)
Cygni, an orange supergiant of
mag. 3.7, 950 l.y. away. *Hale
Observatories photograph.*

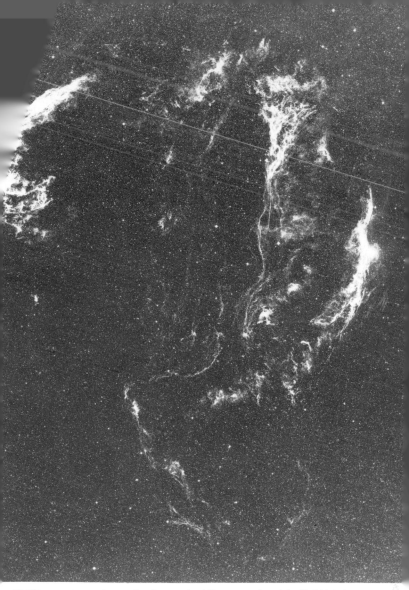

Drifting apart over thousands of years, the delicate traceries of the Veil Nebula are the tenuous remains of a star that shattered itself in a supernova explosion. On the left is the brightest part of the Nebula, NGC 6992, barely visible in large amateur instruments. The fainter arc on the right is NGC 6960, next to the 4th mag. star 52 Cygni. *Hale Observatories photograph.*

DELPHINUS   The Dolphin

A constellation which originated in Greek times, celebrating the lo
standing relationship between mankind and this most intelligent of s
creatures. In legend, dolphins were the messengers of the sea g
Poseidon, and were credited with saving the life of Poseidon's son
Arion, when he was attacked on a ship. Delphinus has a distinctive
shape, its four main stars forming a rectangle known as Job's Coffin.
Its two brightest stars are called Sualocin and Rotanev, which back-
wards read Nicolaus Venator, the Latinized form of the name Niccolo
Cacciatore, who was the assistant to the Italian astronomer Giuseppe
Piazzi. Like its small neighbouring constellations Vulpecula and
Sagitta, Delphinus lies in a rich area of the Milky Way and has become
a famous hunting ground for novae.

α (alpha) Delphini (Sualocin), mag. 3.8, is a blue-white star 170 l.y. away.

β (beta) Del (Rotanev), mag. 3.5, is a white star 110 l.y. away.

γ (gamma) Del, 110 l.y. away, is a showpiece double star consisting of golden and
yellow-white stars of mags. 4.3 and 5.2, neatly separated in small telescopes. In
the same telescopic field of view appears a faint double, Struve 2725, consisting
of stars of 7th and 8th mag.

The Naming of Variable Stars

In addition to the normal system of naming stars in a constellation
(page 6), stars that vary in brightness have their own system of
nomenclature. Stars that were already named when their variability
was discovered, such as δ (delta) Cephei, β (beta) Persei or o (omicron)
Ceti, retain their existing designation. Those apart, variable stars are
denoted by a system of one or two letters or, where that is insufficient,
by the letter V and a number. The first nine variables in a constellation are given the letters R to Z. Then double letters are given, from
RR to RZ. Next the lettering runs from SR to SZ, and so on until ZZ
is reached. Then the sequence goes from AA to AZ, BB to BZ, ending
with QZ, when 334 variable stars have been named. (The letter J is
omitted.) Further variables are given the designation V335, V336 etc.
Novae, too, are given variable star designations. Hence the nova that
erupted in Delphinus in 1967 became known as HR Delphini.

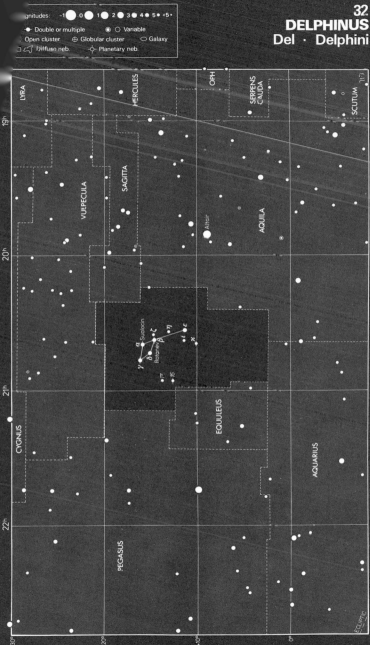

magnitudes: -1 0 1 2 3 4 5 <5

● Double or multiple   ◎ Variable
● Open cluster   ⊕ Globular cluster   ○ Galaxy
Diffuse neb.   ◇ Planetary neb.

LYRA

HERCULES

OPH

SERPENS CAUDA

SCUTUM

VULPECULA

SAGITTA

AQUILA

Altair

CYGNUS

EQUULEUS

AQUARIUS

PEGASUS

19ʰ

20ʰ

21ʰ

22ʰ

+30°

+20°

+10°

0°

Svalocin

α
γ   δ
Rotanev
β   ζ   η   ε
κ
17
15

ECLIPTIC

DORADO   The Goldfish

This constellation, also commonly known as the Swordfish, was introduced in 1603 by Johann Bayer. Its most notable feature is the Large Magellanic Cloud, the bigger of the two satellite galaxies that accompany our Milky Way; the Small Magellanic Cloud lies some 20° away in Tucana.

α (alpha) Doradus, mag. 3.3, is a blue-white star 190 l.y. away.

β (beta) Dor, a yellow supergiant 7500 l.y. away, is one of the brightest Cepheid variable stars. It pulsates between mags. 3.8 and 4.7 every 9 days 20 hrs.

Large Magellanic Cloud. This is a mini galaxy 180,000 l.y. away, a satellite of the Milky Way. Estimates of its mass differ widely, but according to one figure it is about a tenth the mass of our Galaxy and contains perhaps 10,000 million stars. To the naked eye it appears as a fuzzy patch 6° in diameter, 12× the apparent size of the Moon. Binoculars and telescopes show individual stars, nebulae, and clusters.

NGC 2070, the Tarantula Nebula, is a looping cloud of hydrogen gas about 1000 l.y. in diameter in the Large Magellanic Cloud. To the naked eye the nebula appears as a fuzzy star, also known as 30 Doradus. The nebula's popular name, the Tarantula, comes from its spider-like shape. At the centre of the nebula is a cluster of tens of supergiant stars whose light makes the nebula glow; one of these stars is suspected of having a mass of around 1000× that of the Sun. The Tarantula is larger and brighter than any other nebula in the Milky Way. If it were as close as the Orion Nebula, the Tarantula would fill the whole constellation of Orion and would cast shadows.

The elongated shape of the Large Magellanic Cloud, showing many loops of glowing gas in addition to star swarms. The large knot of bright gas (centre left) is the Tarantula Nebula, NGC 2070. *Royal Observatory, Edinburgh.*

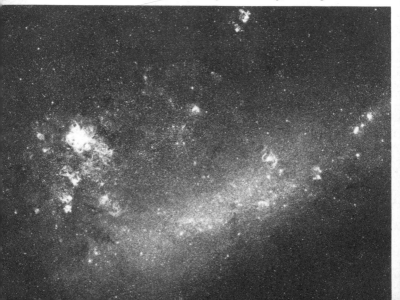

Magnitudes: -1 ⬤ 0 ⬤ 1 ⬤ 2 ⬤ 3 ⬤ 4 ● 5 ● <5 ·
●─● Double or multiple   ◉ ○ Variable
⌒ Open cluster   ⊕ Globular cluster   ⬭ Galaxy
▱ Diffuse neb.   ✦ Planetary neb.

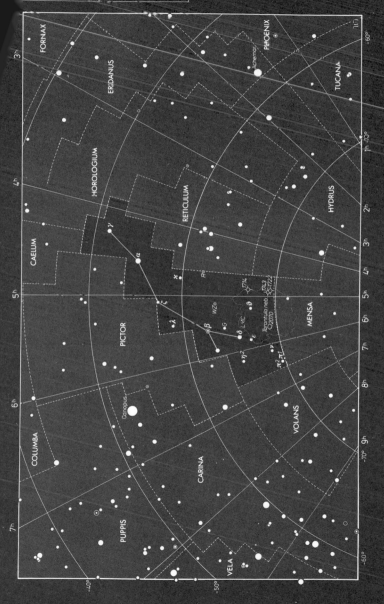

DRACO   The Dragon

Dragons feature in many ancient legends, and so it is not surprising to find such a monster in the sky. Its body lies coiled around the north celestial pole. Perhaps it is the dragon slain by Hercules to obtain the golden apples from the garden of the Hesperides, for one of the feet of Hercules rests upon its head. Although Draco is one of the largest and most ancient constellations, it is indistinct, being comprised of faint stars. Draco contains the pole of the ecliptic, i.e. the point 90° from the plane of the Earth's orbit.

α (alpha) Draconis (Thuban), mag. 3.7, is a white star 230 l.y. away. It was the Pole Star in about 2800 BC, but has lost that place to the present Polaris because of the effect of precession (see page 11).

β (beta) Dra (Rastaban or Alwaid), mag. 2.8, is a yellow supergiant 270 l.y. distant, in the head of Draco.

γ (gamma) Dra (Eltanin, dragon's head), mag. 2.2, is an orange giant 100 l.y. away, and the brightest star in the constellation. From observations of this star, the English astronomer James Bradley discovered the so-called aberration of starlight in 1728.

μ (mu) Dra (Arrakis), 85 l.y. away, is a close double star with cream-coloured components each of mag. 5.8. A telescope of 100 mm aperture or more and high magnification is needed to separate the stars, which orbit every 480 years.

ν (nu) Dra, 120 l.y. away, is a pair of identical white stars of mag. 4.9, easily visible in the smallest of telescopes, and regarded as one of the finest binocular pairs.

o (omicron) Dra, 400 l.y. away, is a mag. 4.7 yellow star with a mag. 8.2 companion visible in small telescopes.

ψ (psi) Dra, 75 l.y. away, is a mag. 4.6 yellow star with a yellow mag. 5.8 companion visible in small telescopes or even binoculars.

16–17 Dra is a wide pair of 5th mag. blue-white stars 330 l.y. away, easily found in binoculars. Telescopes of 60 mm aperture and high magnification show that one of the stars itself has a 6.6 mag. companion, making this a striking triple system.

39 Dra, 170 l.y. away, is another impressive triple system, the two brightest members of which, at mags. 5.0 and 7.1, appear in binoculars as a wide yellow and blue binary. Telescopes of 60 mm aperture or more and high magnification reveal that the brighter star has a mag. 7.7 companion.

NGC 6543 is a 9th mag. planetary nebula 1700 l.y. away, one of the brightest of the class, showing in amateur telescopes an irregular blue-green disk like an out-of-focus star.

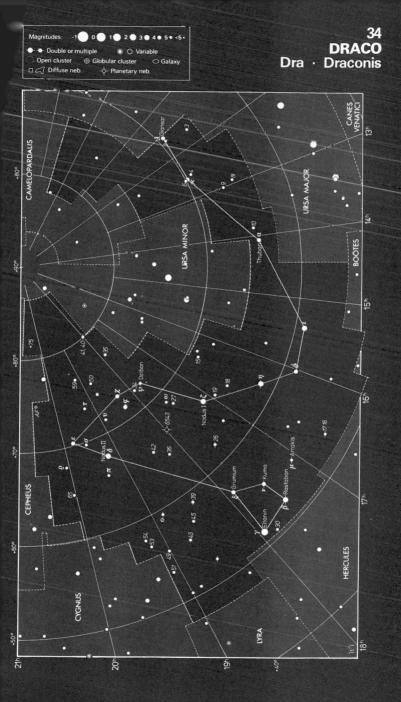

EQUULEUS   The Little Horse

The second-smallest constellation in the sky, Equuleus seems to have
originated with the Greek astronomer Ptolemy in the 2nd Century AD.
It is sometimes also called the Foal or Colt. One myth sees it as Celeris,
the brother of neighbouring Pegasus.

α (alpha) Equulei (Kitalpha, little horse), mag. 3.9, is a yellow giant 150 l.y. away.

γ (gamma) Equ, 180 l.y. away, is a mag. 4.7 white star; a mag. 6 binocular
companion is unrelated.

ε (epsilon) Equ, 130 l.y. away, is a triple star. In small telescopes it appears as a
yellow and blue-white binary of mags. 5.3 and 7.1. But telescopes of 150 mm
aperture show that the brighter (yellow) component is itself made of two close
6th mag. stars; these orbit each other every 101 years.

NGC 1300 in Eridanus, 50 million l.y. away, a classic barred spiral galaxy, with
prominent lanes of dust along its central bar (see page 143). *Hale Observatories
photograph.*

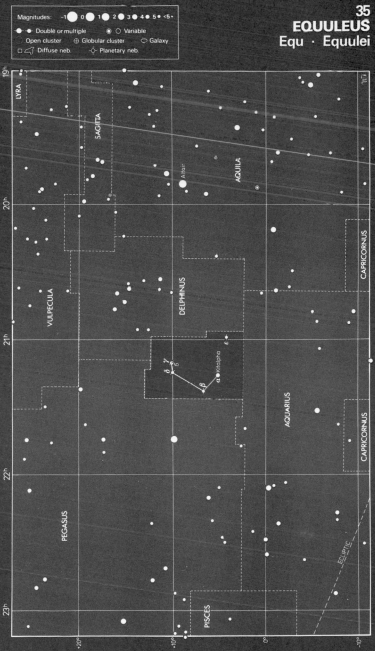

Magnitudes: -1 0 1 2 3 4 5 <5

Double or multiple   Variable
Open cluster   ⊕ Globular cluster   ◯ Galaxy
□ ⌐ Diffuse neb.   ⬦ Planetary neb.

LYRA

SAGITTA

AQUILA

Altair

VULPECULA

DELPHINUS

CAPRICORNUS

δ γ
δ
α Kitalpha
β
ε

AQUARIUS

CAPRICORNUS

PEGASUS

ECLIPTIC

PISCES

ERIDANUS   The River

An extensive constellation, the 6th-largest in the sky, but often over-looked because of its faintness. It meanders from Taurus in the north to Hydrus in the south. In mythology, Eridanus was the river into which Phaethon fell after trying to drive the chariot of his father, the Sun God. But it is also said to represent some real river such as the Po of Italy, the Nile, or the Euphrates. Originally Eridanus included the stars of what is now Fornax, and it stretched only as far as θ (theta) Eridani, which was then known as Achernar, from the Arabic meaning End of the River (its present name, Acamar, comes from Achernar). In recent times Eridanus has been extended to nearly 60° south (below the horizon as visible to the ancient Greeks) and another star has been assigned the title Achernar. The constellation contains several interesting galaxies, all too distant and faint to be picked up easily in amateur telescopes. One celebrated example is NGC 1300, a beautiful 11th magnitude barred spiral pictured on page 140.

α (alpha) Eridani (Achernar, the end of the river), mag. 0.5, is a blue-white star 85 l.y. away.

β (beta) Eri (Cursa, the footstool, referring to its position under the foot of Orion), mag. 2.8, is a blue-white star 91 l.y. away.

ε (epsilon) Eri, mag. 3.7, is one of the most Sun-like of the nearby stars, lying 10.7 l.y. away. A yellow dwarf, it is believed to be accompanied by a large planet or a very small star. ε (epsilon) Eridani has been the subject of several attempts to listen for interstellar radio messages, but nothing has been heard.

θ (theta) Eri (Acamar), 55 l.y. away, is a striking pair of blue-white stars of mags. 3.2 and 4.4 visible in small telescopes.

o^2 (omicron^2) Eri, 15.9 l.y. away, also known as 40 Eridani, is a remarkable triple star. A small telescope shows that the mag. 4.4 yellow primary, a star similar to the Sun, has a wide mag. 9.6 white dwarf companion, the most easily seen white dwarf in the sky. Telescopes of 75 mm aperture or above reveal that the white dwarf has an 11th mag. companion, which is a red dwarf, thereby completing a most interesting trio.

32 Eri, 220 l.y. away, is a beautiful double star for small telescopes, consisting of a mag. 5 yellow star and a blue-green mag. 6.3 companion.

NGC 1535 is a small 9th mag. planetary nebula 2150 l.y. away. Small telescopes show it, but apertures of 150 mm are needed to appreciate its blue-grey disk.

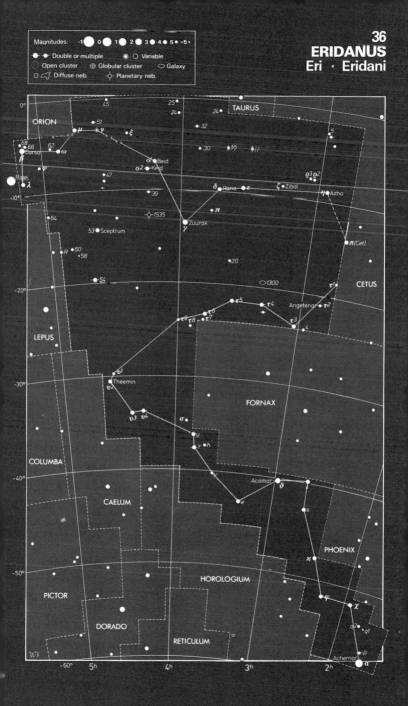

Magnitudes: -1 0 1 2 3 4 5 <5
●—● Double or multiple   ⊙ Variable
◯ Open cluster   ⊕ Globular cluster   ◯ Galaxy
☐ Diffuse neb.   ✧ Planetary neb.

ORION
TAURUS
CETUS
LEPUS
FORNAX
COLUMBA
CAELUM
PHOENIX
PICTOR
HOROLOGIUM
DORADO
RETICULUM

μ
ν
ξ
51
45
25
24
24
32
5
68 66
62
ω
Cursa
ψ
Rigel
λ
o¹ Beid
o² Keid
47
39
δ Rana ε
ζ Zibal
η Azha
θ³ θ²
π
γ Zaurak
53 ● Sceptrum
64
R 60
58
54
20
1535
1300
τ⁵
τ⁴
τ⁶
τ⁹ τ⁸ τ⁷
Angetenar τ²
τ³
λ
τ¹
ν¹
Theemin
ν²
ν³
ν⁴
σ
y
f ● h
χ
Acamar θ
e
s
κ
φ
χ
a² q¹
ρ
Achernar
α

0°
-10°
-20°
-30°
-40°
-50°
-60°

5ʰ   4ʰ   3ʰ   2ʰ

FORNAX   The Furnace

A barren constellation introduced in the mid–18th Century by Nicolas Louis de Lacaille, originally under the name of Fornax Chemica, the Chemical Furnace. Unknown to Lacaille, the constellation contains a dwarf elliptical galaxy of our Local Group, approximately 800,000 light years from the Milky Way, but too faint to see in amateur telescopes. A 13th mag. globular cluster, NGC 1049, is part of the dwarf galaxy. Fornax also contains a compact cluster of galaxies about 100 million l.y. away, on the border with Eridanus.

α (alpha) Fornacis, 46 l.y. away, is a yellow star of mag. 4.1 with a close mag. 6.6 companion.

β (beta) For, mag. 4.5, is a yellow giant star 200 l.y. away.

Among the many distant galaxies in Fornax is this 11th mag. barred spiral, NGC 1097, 40 million l.y. away, located near the centre of the constellation. *Anglo-Australian Telescope Board.*

Magnitudes: -1 0 1 2 3 4 5 <5
● ● Double or multiple   ○ Variable
○ Open cluster   ⊕ Globular cluster   ○ Galaxy
□ ◻ Diffuse neb.   ◇ Planetary neb.

CETUS

PHOENIX

SCULPTOR

2ʰ

χ
π
μ
φ
x
ω
η³ η²
γ²
β
ζ
α
χ²
ϱ
δ

3ʰ

HOROLOGIUM

DORADO

ERIDANUS

CAELUM

PICTOR

4ʰ

COLUMBA

LEPUS

5ʰ

-10°

-20°

-30°

-50°

6ʰ

GEMINI   The Twins

A constellation familiar since ancient times, representing a pair of twins holding hands. We know them as Castor and Pollux, members of the Argonauts' crew, and of mixed parentage: although both sons of Leda, Castor's father was King Tyndareus of Sparta, while the father of Pollux was the god Zeus. The twins were patron saints of mariners, appearing in ships' rigging as the electrical phenomenon known as Saint Elmo's Fire. Castor and Pollux provide a useful yardstick in the sky for measuring angular distances: they are 4.5° apart. Gemini is a major zodiacal figure, the Sun passing through the constellation from late June to late July. Each year the Geminid meteors radiate from a point near Castor. The Geminids are one of the year's richest and most brilliant showers, reaching a maximum around December 14, when up to 60 meteors per hour may be seen.

α (alpha) Geminorum (Castor), 45 l.y. away, is an astounding multiple star, consisting of six separate components. To the naked eye it appears as a blue-white star of mag. 1.6. A 60 mm telescope with high magnification splits Castor into two close components, of mags. 2.0 and 2.9, which orbit each other every 420 years. Separation between these stars is increasing, so they will become increasingly easy to split until about the year 2065, when they start to close up again. Both these stars are spectroscopic binaries. Small telescopes also show a wide 9th mag. red dwarf companion to Castor; this is itself an eclipsing binary, completing the 6 star system.

β (beta) Gem (Pollux), mag. 1.1, is the brightest star in the constellation. Some astronomers suggest that Pollux, being called β (beta) Geminorum, was originally fainter than Castor and has since brightened or Castor has faded. But the truth is that Johann Bayer, who labelled the stars with Greek letters in 1603, did not distinguish carefully which star was the brighter, and so caused unnecessary confusion. Pollux is an orange giant 36 l.y. away.

γ (gamma) Gem, mag. 1.9, is a blue-white star 85 l.y. away.

δ (delta) Gem, 59 l.y. away, is a creamy white star of mag. 3.5 with an orange dwarf companion of mag. 8.1. The brightness contrast makes the pair difficult in telescopes below about 75 mm aperture.

ε (epsilon) Gem, mag. 3.0, is a yellow supergiant 680 l.y. away. Powerful binoculars, or a small telescope, reveal a wide 9th mag. companion.

ζ (zeta) Gem, 1400 l.y. away, is both a variable star and a binocular double. A yellow supergiant, it is a Cepheid variable, fluctuating between mags. 3.7 and 4.1 every 10.2 days. Binoculars or small telescopes reveal a wide 8th mag. companion, which is unrelated.

η (eta) Gem, 190 l.y. away, is another double variable. A red giant star, it fluctuates in semi-regular manner between mags. 3.1–3.9 about every 230 days. It also has a close 8th mag. companion, requiring a large telescope to distinguish it from the glare of the primary. ▶

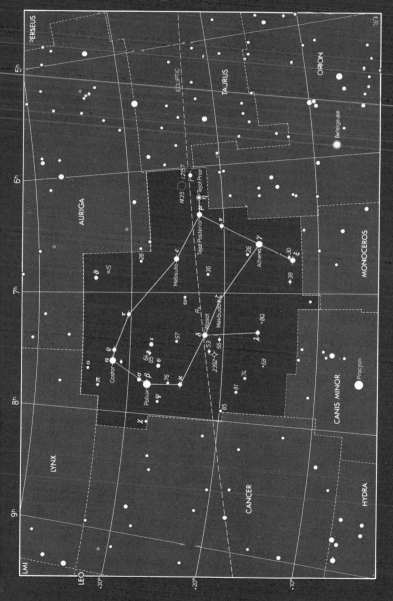

GRUS The Crane

A constellation introduced on the 1603 star atlas of Johann Bayer. It represents a water bird, the Crane, although others have seen here a flamingo. The stars δ (delta) and μ (mu) Gruis are striking naked-eye doubles.

α (alpha) Gruis (Alnair, the bright one), mag. 1.7, is a blue-white star 91 l.y. away.

β (beta) Gru, mag. 2.1, is a red giant 270 l.y. away.

γ (gamma) Gru, mag. 3.0, is a blue-white star 230 l.y. away.

δ (delta) Gru is a naked eye double star; the two components are believed to be at different distances and are therefore unrelated. δ^1 (delta[1]), of mag. 4.0, is a yellow giant 140 l.y. away. δ^2 (delta[2]) is a mag. 4.1 red star of uncertain distance.

μ (mu) Gru is a pair of naked eye stars which appear in the same line of sight by chance. μ^1 (mu[1]) is a yellow giant of mag. 4.8, 260 l.y. away. μ^2 (mu[2]) is a mag. 5.1 yellow giant 280 l.y. away.

◀ κ (kappa) Gem, 150 l.y. away, is a mag. 3.6 yellow giant, with a mag. 9.5 companion made difficult in small telescopes because of the extreme brightness difference.

ν (nu) Gem, mag. 4.2, is a blue-white star 360 l.y. away with a wide 9th mag. companion visible in binoculars or small telescopes.

38 Gem, 85 l.y. away, is a double star for small telescopes, with white and yellow components of mags. 4.7 and 7.6.

M 35 (NGC 2168) is a large, bright star cluster visible to the naked eye or in binoculars, covering an area of sky greater than the Moon. It consists of about 120 stars, 2600 l.y. away. Even small telescopes at low magnification show the stars of this outstanding cluster to be arranged in curving chains. Nearby in the sky, but actually about 14,000 l.y. farther off, is NGC 2158, a very rich star cluster that appears as a small, faint patch of light requiring at least 100 mm aperture to distinguish.

NGC 2392 is an 8th mag. planetary nebula sometimes known as the Eskimo or Clown Face Nebula because of its appearance in large telescopes. Small telescopes show it as a fairly large, rounded, blue-green disk. Its central star is of 10th mag. NGC 2392 lies about 1400 l.y. away.

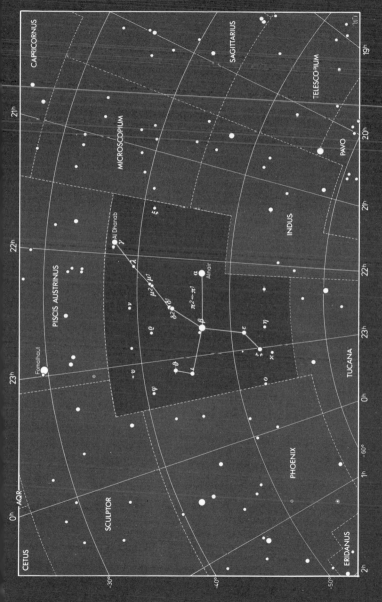

HERCULES

Hercules represents a Greek mythological hero, famous for his twelve labours. Earlier visualizations were of an anonymous kneeling man, with one foot on the celestial Dragon, Draco, which lies adjacent. Some legends identify the constellation with the ancient Sumerian superman, Gilgamesh. Hercules is not one of the most prominent constellations, despite being 5th-largest in the sky. But it is stocked with abundant double stars for users of small telescopes, plus one of the brightest and richest globular clusters in the sky, M 13, easily located on one side of the central 'keystone' of four stars which marks the pelvis of Hercules.

α (alpha) Herculis (Ras Algethi, kneeler's head), 540 l.y. distant, is a red supergiant star about $600\times$ the Sun's diameter, making it one of the largest stars known. Like most red giants it is erratically variable, in this case fluctuating from about mag. 3 to mag. 4. It is actually a double star, with a mag. 5.4 blue-green companion visible in small telescopes.

β (beta) Her, mag. 2.8, is a yellow giant 100 l.y. away.

γ (gamma) Her, mag. 3.8, is a white star 140 l.y. away, with a wide 10th mag. companion visible in small telescopes.

δ (delta) Her, mag. 3.1, is a white star 91 l.y. away. Small telescopes show a mag. 8.8 star nearby, which is physically unrelated, making this an optical double.

ζ (zeta) Her, 31 l.y. away, is a mag. 3.1 yellow star with a close mag. 5.6 red companion which orbits it every 34 years. They reach maximum separation in late 1990, but even then will require 100 mm aperture to be split.

κ (kappa) Her, 280 l.y. away, is a mag. 5.0 yellow giant with a mag. 6.3 companion easily seen in small telescopes.

ϱ (rho) Her, 170 l.y. away, is a blue-white star of mag. 4.5 with a mag. 5.5 companion visible in small telescopes.

95 Her, 470 l.y. away, is a double star for small telescopes, with components of mags. 5.2 and 5.1, appearing gold and silver.

M 13 (NGC 6205) is a globular cluster of 300,000 stars, the brightest of its kind in the northern skies. It can be seen by the naked eye, and is unmistakable in binoculars. The cluster lies 22,500 l.y. away, and is at least 100 l.y. in diameter. Small telescopes resolve individual stars throughout the cluster, giving a mottled, sparkling effect.

M 92 (NGC 6341) is a globular cluster only slightly inferior to its more famous neighbour, M 13, which overshadows it. M 92 is easily seen in binoculars. It is more condensed at the centre than M 13, and needs a larger telescope to resolve its stars. It lies 36,000 l.y. away.

NGC 6210 is a 10th mag. planetary nebula, which telescopes 75 mm or larger show as a blue-green ellipse.

HOROLOGIUM   The Pendulum Clock

One of the constellations representing mechanical instruments introduced in the 1750s by the Frenchman Lacaille. As with so many of his
constellations, Horologium is faint and obscure.

α (alpha) Horologii, mag. 3.9, is a yellow giant star 190 l.y. away.

β (beta) Hor, mag. 5.0, is a white star 280 l.y. away.

NGC 1512 is a barred spiral galaxy 50 million l.y. away in Horologium with a
smaller, 12th mag. elliptical companion, NGC 1510. *Anglo-Australian Telescope
Board.*

HYDRA   The Water Snake

The largest constellation in the sky, but by no means easy to identify on account of its faintness. Apart from its brightest star, Alphard, which marks the heart of the Water Snake, Hydra's only readily recognizable feature is its head, made up of an attractive group of six stars. Hydra winds its way from the head in the northern celestial hemisphere on the borders of Cancer to the tip of its tail south of the celestial equator adjacent to Libra and Centaurus. Its total length is over 100°. Various peoples have seen the figure as a Serpent, a Snake, or even a Dragon, a characterization now reserved for Draco. Hydra is usually identified with the multiheaded monster slain by Hercules. Another legend links it with Corvus the Crow and Crater the Cup which are found on its back, the bird having returned to the god Apollo with Hydra in its claws as an excuse for its delayed mission to fetch water in the cup.

α (alpha) Hydrae (Alphard, the solitary one), mag. 2.0, is an orange giant 130 l.y. away.

β (beta) Hya, mag. 4.3, is a blue-white star 270 l.y. away.

γ (gamma) Hya, mag. 3.0, is a yellow giant 105 l.y. away.

δ (delta) Hya, mag. 4.2, is a blue-white star 140 l.y. away.

ε (epsilon) Hya, 110 l.y. away, is a beautiful but difficult double star of contrasting colours, needing high power on telescopes of at least 75 mm aperture. The stars are yellow and blue, of mags. 3.5 and 6.9.

54 Hya, 150 l.y. away, is an easy double for small telescopes, consisting of yellow and purple stars of mags. 5.2 and 7.1.

R Hya, 330 l.y. away, is a red giant variable similar to Mira in Cetus. It fluctuates between mags. 4–10 every 386 days.

U Hya is a deep red variable star that fluctuates irregularly between mags. 4.8–5.8. Its distance is uncertain.

M 48 (NGC 2548) is a cluster of about 80 stars, 3000 l.y. away, just visible to the naked eye under clear skies but easily seen in binoculars. It is somewhat triangular in shape, covering an area of sky equivalent to the apparent size of the Moon, and is well shown in amateur telescopes under low power.

M 83 (NGC 5236) is a large, face-on spiral galaxy of 8th mag., visible in small telescopes. It has a bright nucleus, and its spiral arms can be traced in apertures of 150 mm. (See photograph on page 156.)

NGC 3242 is a 9th mag. planetary nebula of similar apparent size to the disk of Jupiter; in fact it is popularly termed the 'ghost of Jupiter'. This often-overlooked object 1900 l.y. away is prominent enough to be picked up in small telescopes at low magnification as a blue-green disk, while larger instruments show a bright inner disk surrounded by a fainter halo.

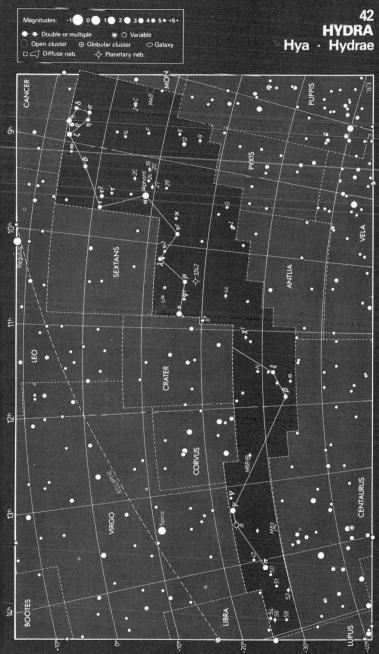

HYDRUS   The Lesser Water Snake

Johann Bayer, the German celestial cartographer, introduced this constellation in 1603 as a smaller southern counterpart of the great Water Snake, Hydra. Bayer's Hydrus, sandwiched between the two Magellanic Clouds, almost bridges the gap between Eridanus and the south celestial pole. Alas, there is little to interest the casual observer.

α (alpha) Hydri, mag. 2.9, is a white star 36 l.y. away.

β (beta) Hyi, mag. 2.8, is a yellow star 21 l.y. away.

γ (gamma) Hyi, mag. 3.2, is a red giant star 160 l.y. away.

π (pi) Hyi is a binocular pair of mag. 5.5 red and orange stars, unrelated. π^1 (pi^1) lies 400 l.y. away; π^2 (pi^2) is 520 l.y. away.

The face-on spiral galaxy M 83 in Hydra, a turbulent swirl of stars and gas. For description, see page 154. *Anglo-Australian Telescope Board.*

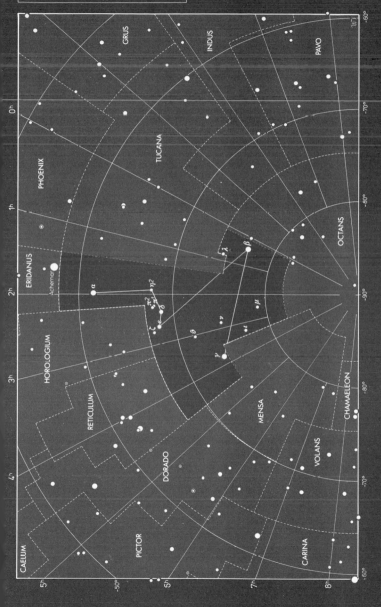

INDUS   The Indian

A constellation representing an American native Indian, introduced in 1603 on the star atlas of Johann Bayer. It has no stars brighter than 3rd magnitude.

α (alpha) Indi, mag. 3.1, is an orange giant 120 l.y. away.

β (beta) Ind, mag. 3.7, is an orange giant 270 l.y. away.

δ (delta) Ind, mag. 4.4, is a white star 110 l.y. away.

ε (epsilon) Ind, mag. 4.7, is a yellow dwarf star similar to the Sun but slightly smaller and cooler, lying 11.2 l.y. away, making it one of the Sun's closest neighbours.

θ (theta) Ind, 91 l.y. away, is a pair of stars of mags. 4.6 and 7.0, divisible in small telescopes.

Between Indus and Hydrus lies the Small Magellanic Cloud, an irregularly shaped splash of stars. To the right of it in this photograph is the globular cluster 47 Tucanae, which lies in our own Galaxy. For descriptions, see page 240. *Royal Observatory Edinburgh.*

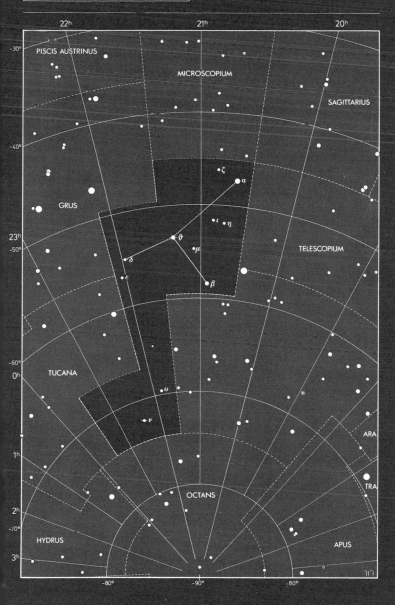

LACERTA   The Lizard

An inconspicuous constellation sandwiched between Cygnus and Andromeda, introduced in the late 17th Century by the Polish astronomer Johannes Hevelius. An alternative constellation in this area was the Sceptre and Hand of Justice created in 1697 by the Frenchman Augustin Royer to commemorate King Louis XIV; the German Johann Elert Bode in 1787 called this region Frederick's Glory in honour of King Frederick II of Prussia. Both these alternatives have been discarded in favour of Lacerta. The constellation's most celebrated object is BL Lacertae, originally thought to be a peculiar 14th magnitude variable star. It is now known as the prototype of a group of objects believed to be giant elliptical galaxies, with variable centres, lying far off in the Universe and evidently related to quasars. Three bright novae have occurred within the boundaries of Lacerta this century.

α (alpha) Lacertae, mag. 3.8, is a blue-white star 98 l.y. away.

β (beta) Lac, mag. 4.4, is a yellow giant 220 l.y. away.

Johannes Hevelius (1611–1687)

Johannes Hevelius of Danzig was one of the finest observers of his day. His masterwork was a catalogue of 1564 stars, published posthumously in 1690. The catalogue was accompanied by a set of sky maps, beautifully engraved by Hevelius himself, on which he introduced seven constellations still in use today: Canes Venatici, Lacerta, Leo Minor, Lynx, Scutum, Sextans and Vulpecula. Another important cartographic product of Hevelius was his map of the Moon, published in 1647 in his book *Selenographia*. It was the first major Moon map to be made, and introduced the first system of lunar nomenclature. He named lunar formations after features on Earth: for instance, the crater Copernicus he called Etna; Tycho was Mount Sinai; and Mare Imbrium was the Mediterranean. A few of the names given by Hevelius remain, such as the lunar Alps and Apennines; but his names have mostly been replaced by the system of Giovanni Battista Riccioli (1598–1671), who named the craters after famous philosophers and astronomers rather than places on Earth. Fittingly enough, the craters commemorating Hevelius and Riccioli are to be found close together on the Moon.

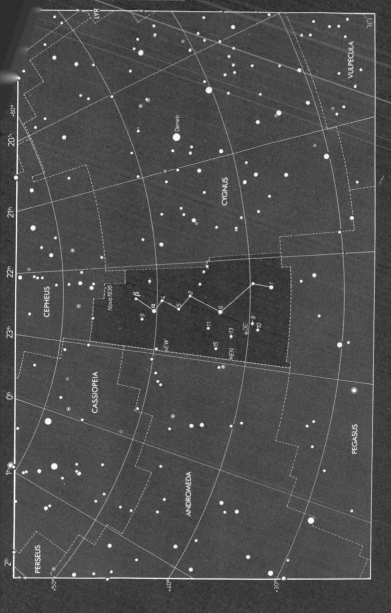

-1 0 1 2 3 4 5 <5

ble or multiple · ○ Variable
luster ⊕ Globular cluster ○ Galaxy
ffuse neb. ◇ Planetary neb.

LYR

VULPECULA

+60°

20ʰ

Deneb

CYGNUS

21ʰ

22ʰ

CEPHEUS

Nova1936

β
α 4
9 5 2 1
11 6
EW DG 8
15 EN 13 10

23ʰ

0ʰ

CASSIOPEIA

PEGASUS

1ʰ

ANDROMEDA

2ʰ

PERSEUS

+50° +40° +30°

LEO The Lion

One of the few constellations that looks like the figure it is suppo[se]
represent – in this case, a crouching lion. The Lion's head is ou[t]
by a sickle shape of six stars from ε (epsilon) to α (alpha) Leonis,
the Lion's body stretching out behind, its tail marked by β (b[eta])
Leonis. This is the Lion reputedly slain by the hero Hercules as t[he]
first of his 12 labours. Leo is a large and bright constellation, containin[g]
many stars and galaxies of interest. The Sun passes through the constellation from mid-August to mid-September. Every year, around
November 17, the Leonid meteors appear to radiate from a point near
γ (gamma) Leonis. Usually the numbers seen are low, peaking at about
10 per hour, but on occasion spectacular storms of meteors have been
recorded, when as many as 100,000 can fall per hour, like celestial
snowflakes. The last such Leonid meteor storm was seen from the
United States in 1966. Leo contains a little-known star, Wolf 359,
which is the 3rd-nearest star to the Sun. It is a red dwarf 7.6 l.y. away,
appearing of magnitude 13.5; it is located in the south of the constellation, near the border with Sextans. Leo contains numerous distant
galaxies, the brightest of which are mentioned below, plus two faint
dwarf members of our Local Group, beyond the reach of amateur
telescopes.

α (alpha) Leonis (Regulus), mag. 1.4, is a blue-white star 85 l.y. away. It has a
wide companion of mag. 7.6, visible in binoculars or small telescopes.

β (beta) Leo (Denebola, lion's tail), mag. 2.1, is a white star 42 l.y. away.

γ (gamma) Leo (Algieba, lion's mane), 100 l.y. away, consists of a pair of golden-
yellow giant stars of mags. 2.3 and 3.5, orbiting every 620 years. They form an ▶

Trails of over 20 Leonid meteors
swept across the stars of the Big
Dipper in this 43-second
exposure, made during the great
Leonid meteor storm of 1966.
*Photograph by Dave McLean,
Kitt Peak.*

LEO MINOR   The Lesser Lion

Leo Minor lies between the larger and brighter Leo, and Ursa
Johannes Hevelius, the Polish astronomer, introduced this conste
in the late 17th Century. There is little of interest in Leo Minor, ar
labelling of its stars is very fragmentary, testimony to cavalier treatr
by successive generations of celestial cartographers.

β (beta) Leo Minoris, mag. 4.2, is an orange giant star 100 l.y. away.

46 LMi, mag. 3.8, is an orange giant 75 l.y. away.

◀ exceptionally handsome double in small telescopes, one of the finest in the sky.
In binoculars, an unrelated 5th mag. yellow star is visible nearby.

δ (delta) Leo (Zosma), mag. 2.6, is a blue–white star 52 l.y. away.

ε (epsilon) Leo, mag. 3.0, is a yellow giant 310 l.y. away.

ζ (zeta) Leo, mag. 3.4, is a white star 120 l.y. away. Binoculars show an unrelated
orange background star nearby, of mag. 5.8. A 3rd star, mag. 6.0 and also
unrelated, can be seen farther off in binoculars, making this an optical triple.

ι (iota) Leo, 78 l.y. away, is a close and difficult double star. It shines to the naked
eye as a yellow–white star of mag. 4.0, but actually consists of components of
mags. 4.1 and 7.0 orbiting every 200 years. The stars are at present moving apart,
and apertures of 150 mm should be sufficient to separate them.

R Leo is a red giant variable of the Mira type, lying more than 3000 l.y. away. It
appears strongly red when at maximum. R Leonis varies between mags. 5.4 and
10.5 every 313 days.

M 65, M 66 (NGC 3623, NGC 3627) are a pair of spiral galaxies lying 20 million
l.y. away. At 9th and 8th mag. respectively they can be detected in large binoculars under clear conditions, but at least 100 mm aperture at low power is required
for their elongated shape and condensed centres to be clearly seen.

M 95, M 96 (NGC 3351, NGC 3368) are a pair of spiral galaxies of 10th and 9th
mag. respectively, 22 million l.y. away, visible as circular nebulosities in small
telescopes. About 1° away lies the smaller M 105 (NGC 3379), a 9th mag.
elliptical galaxy.

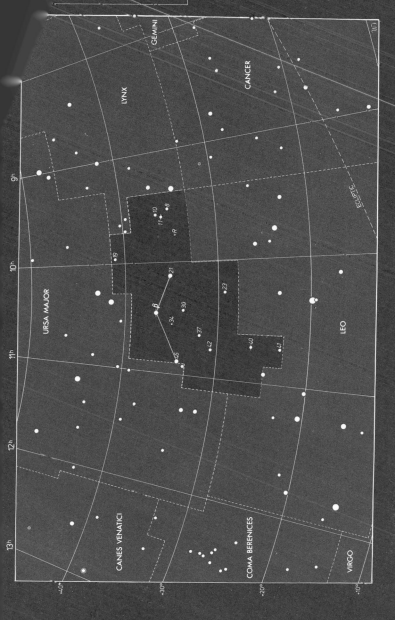

LEPUS The Hare

Lepus is a constellation known since ancient Greek times. It rep‑
a Hare, cunningly located at the feet of its Hunter, Orion. The F.
also associated in many legends with the Moon. For instance, the
liar figure of the Man-in-the-Moon is sometimes interpreted as a
or rabbit. So perhaps Lepus is another incarnation of the lunar ha.
Lepus is overshadowed by Orion's brilliance, but is not without intere.
for amateur observers.

α (alpha) Leporis (Arneb, hare), mag. 2.6, is a yellow-white supergiant star, 950 l.y. away.

β (beta) Lep, mag. 2.8, is a yellow giant star 320 l.y. away.

γ (gamma) Lep, 27 l.y. away, is an attractive binocular duo consisting of a yellow star of mag. 3.8 with an orange companion of mag. 6.4.

δ (delta) Lep, mag. 3.8, is a yellow giant 160 l.y. away.

ε (epsilon) Lep, mag. 3.2, is an orange giant 160 l.y. away.

κ (kappa) Lep, 420 l.y. away, is a mag. 4.5 blue-white star with a close mag. 7.5 companion, difficult to see in the smallest telescopes because of the magnitude contrast.

R Lep is an intensely red star known as Hind's Crimson Star after the English observer John Russell Hind who described it in 1845 as 'like a drop of blood on a black field'. R Leporis varies every 430 days or so between 6th and 10th mag.

M 79 (NGC 1904) is a small but rich globular cluster 43,000 l.y. away, visible as a fuzzy 8th mag. star in small telescopes. Nearby in the same low-power field is the multiple star Herschel 3752, consisting of a mag. 5.5 primary with two companions, a wide one of 9th mag. and a close one of mag. 6.7, all visible in small telescopes.

NGC 2017 is a small but remarkable star cluster, also known as the multiple star Herschel 3780. Amateur telescopes reveal a group of five stars ranging from 7th to 10th mag. Two of the stars are close binaries requiring at least 100 mm aperture to split, so this is actually a family of at least seven related stars.

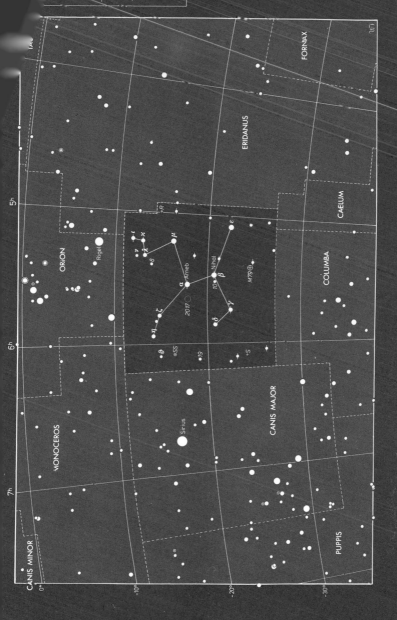

LIBRA   The Scales

A small faint constellation of the zodiac, through which the Sun p.
during November. The ancient Greeks knew it as the 'Claws of
Scorpion', an extension of neighbouring Scorpius, but the Romans ma
it into a separate constellation in the time of Julius Caesar. Since the
the Scales of Libra have come to be regarded as the symbol of justice
held aloft by the goddess of justice, Astraea. One legend identifies
Astraea with the neighbouring figure of Virgo. Although faint, Libra
contains several interesting stars.

α (alpha) Librae (Zubenelgenubi, the southern claw), 72 l.y. away, is a wide
binocular double consisting of a blue-white star of mag. 2.8 with a white
companion of mag. 5.2.

β (beta) Lib (Zubeneschamali, northern claw), mag. 2.6, is celebrated as one of
the few bright stars to show a distinct greenish tinge. It lies 120 l.y. away.

γ (gamma) Lib (Zubenelakrab, scorpion's claw), mag. 3.9, is a yellow giant 75
l.y. away.

δ (delta) Lib, 240 l.y. away, is an eclipsing variable of the Algol type. It varies
between mags. 4.8 and 5.9 every 2 days 8 hrs.

ι (iota) Lib is a multiple star 300 l.y. away. Its main blue-white component, of
mag. 4.5, has a wide 10th mag. companion which is a difficult object to see in the
smallest telescopes because of the brightness difference. Under high power, telescopes of 75 mm aperture and above show this fainter companion itself to consist
of two matched 10th mag. stars. Binoculars show a 6th mag. star nearby called 25
Librae; this star may also be associated with the system.

μ (mu) Lib, 300 l.y. away, is a close double star consisting of components of mags.
5.8 and 6.7, divisible in telescopes of 75 mm aperture or over.

48 Lib, mag. 4.9, is a *shell star* similar to γ (gamma) Cassiopeiae and Pleione in
Taurus. It is a blue giant with an abnormally high speed of rotation that causes
it to throw off a ring of gas from its equator.

NGC 5897 is a large but loosely scattered and faint globular cluster 45,000 l.y.
away, unspectacular in small instruments.

Magnitudes: -1 ● 0 ● 1 ● 2 ● 3 ● 4 ● 5 ● <5 ·
●—● Double or multiple ◉ ○ Variable
Open cluster ⊕ Globular cluster ○ Galaxy
Diffuse neb. ✦ Planetary neb.

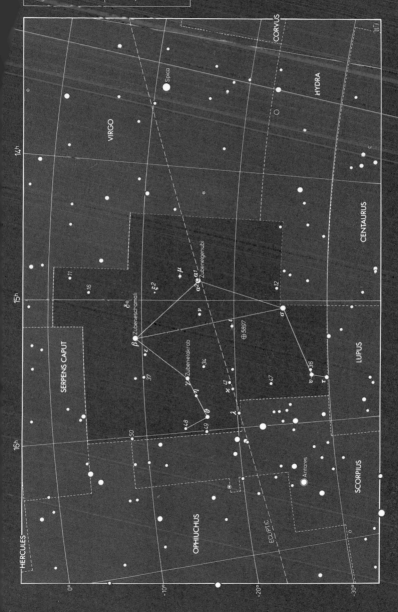

LUPUS   The Wolf

Lupus is stocked with numerous interesting objects, although it is often overlooked in favour of its more spectacular neighbours Scorpius and Centaurus. The constellation was regarded by the Greeks and Romans as an unspecified wild animal, held in the grasp of Centaurus as an offering to the gods. Its identification as a wolf seems to have become common in Renaissance times. Lupus lies in the Milky Way, and is rich in double stars.

α (alpha) Lupi, mag. 2.3, is a blue giant star 680 l.y. away.

β (beta) Lup, mag. 2.7, is a blue-white star 360 l.y. away.

γ (gamma) Lup, mag. 2.8, is a blue-white star 260 l.y. away.

ε (epsilon) Lup, 460 l.y. away, is a multiple star, consisting of a blue-white primary of mag. 3.4 with a wide 9th mag. companion visible in small telescopes. Telescopes of 220 mm aperture and above, with high magnification, reveal that the brighter star also has a close mag. 6 companion.

η (eta) Lup, 490 l.y. away, is a double star consisting of a mag. 3.4 blue-white primary with a mag. 7.7 companion – not easy to see in small telescopes because of the magnitude contrast.

κ (kappa) Lup, 150 l.y. away, is an easy double star for small telescopes, consisting of blue-white components of mags. 3.7 and 5.7.

μ (mu) Lup, 250 l.y. away, is a multiple star. Small telescopes reveal a mag. 4.3 blue-white primary with a wide 7th mag. companion. But in telescopes of 100 mm and above, with high magnification, the primary itself is seen to be double, consisting of two 5th mag. stars.

ξ (xi) Lup, 160 l.y. away, is a neat pair of mag. 5.2 and 5.6 white stars, well seen in small telescopes.

π (pi) Lup, 420 l.y. away, appears to the naked eye as mag. 3.9, but telescopes above 75 mm aperture show that it consists of two close blue-white stars of mags. 4.7 and 4.8.

NGC 5822 is a large, loose cluster of about 120 stars, 6000 l.y. away, visible in binoculars or small telescopes.

NGC 5986 is a small 9th mag. globular cluster, 45,000 l.y. distant, visible as a rounded patch in small telescopes.

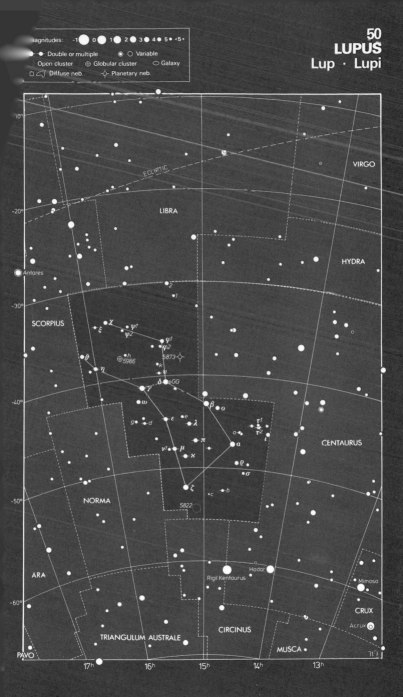

LYNX

A decidedly obscure constellation, despite its considerable size (larger than Gemini, for instance). It was introduced in the late 17th Century by the Polish astronomer Johannes Hevelius to fill the gap between Ursa Major and Auriga; he named it Lynx, it is said, because only the lynx-eyed would be able to examine it. But despite its faintness, owners of small telescopes will find many exquisite double stars within it.

α (alpha) Lyncis, mag. 3.1, is a red giant 170 l.y. away.

5 Lyn, 330 l.y. away, is an orange giant star of mag. 5.2 with an unrelated, wide 8th mag. companion, visible in small telescopes.

12 Lyn, 180 l.y. away, is a fascinating triple star. Small telescopes show a mag. 4.9 blue-white star with a mag. 8.5 companion. Telescopes of 75 mm aperture and over reveal that the brighter component is itself a close double, of mags. 5.4 and 6.0.

15 Lyn, 105 l.y. away, is a close double star for telescopes of 150 mm aperture and above. The stars are of mags. 4.9 and 6.1, the brighter star appearing a deep yellow colour.

19 Lyn, 420 l.y. away, is an attractive double star for small telescopes, with components of mags. 5.6 and 6.5. There is also a very wide 8th mag. star that may be related.

38 Lyn, 88 l.y. away, is a pair of 4th and 6th mag. stars, difficult in the smallest telescopes because of their closeness.

41 Lyn, 68 l.y. away, is actually over the border in Ursa Major. It is a yellow star of mag. 5.4, and small telescopes reveal a wide companion of mag. 7.8. A 10th mag. star nearby forms a triangle, making this an apparent triple.

Flamsteed Numbers

Apart from α (alpha) Lyncis, all the named stars in Lynx are referred to not by Greek letters but by Flamsteed numbers. These numbers stem from a catalogue of 2935 stars, *Historia Coelestis Britannica*, compiled by the first British Astronomer Royal, John Flamsteed (1646–1719). The catalogue was published posthumously in 1725. Flamsteed listed the stars in each constellation in order of right ascension. The numbers that are now known as Flamsteed numbers were not given by him, but were added later by other astronomers.

Magnitudes: −1 ● 0 ● 1 ● 2 ● 3 ● 4 ● 5 ● <5 •
● ● Double or multiple   ◉ ○ Variable
◯ Open cluster   ⊕ Globular cluster   ◯ Galaxy
▢ ◁ Diffuse neb.   ◇ Planetary neb.

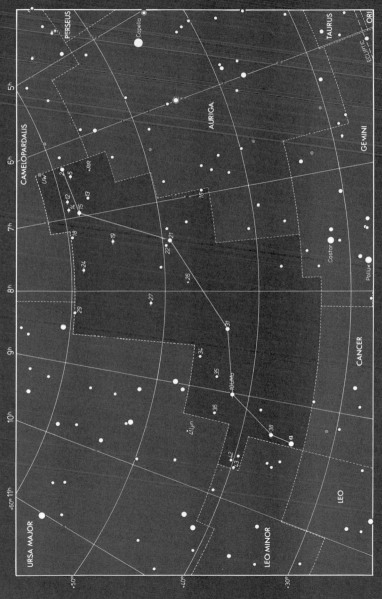

LYRA   The Lyre

A constellation dating from ancient times, representing the stringed instrument invented by Hermes and given by his half-brother Apollo to Orpheus. This constellation has also been visualized as an eagle or vulture. Although small, Lyra is bright and prominent. Lyra contains the 5th-brightest star in the sky, Vega, which forms one corner of the famous Summer Triangle (Deneb in Cygnus and Altair in Aquila mark the other corners). Our Sun's motion around the Galaxy is carrying it and the Solar System in the direction of Vega at a velocity of 20 km per sec. relative to nearby stars. Because of precession, Vega will become the Pole Star in about 14,000 AD. Two meteor showers emanate from Lyra, one in April and one in June. The April Lyrids, which reach maximum around April 21–22, are the more abundant, up to 12 per hour being visible. The June Lyrids peak at about 8 meteors per hour on June 16 each year.

α (alpha) Lyrae (Vega), mag. 0.03, is a brilliant blue-white star 26 l.y. away. It is the 5th-brightest star in the sky.

β (beta) Lyr (Sheliak), 1100 l.y. away, is a remarkable multiple star. Small telescopes easily resolve it as a double star of cream and blue components. The fainter, blue star is of mag. 7.8; the brighter star is an eclipsing binary that varies between mags. 3.4 and 4.3 every 12.9 days. These eclipsing stars are so close together that gravity distorts them into egg shapes, and hot gas spirals off them into space.

γ (gamma) Lyr, mag. 3.2, is a blue-white star 190 l.y. away. ▶

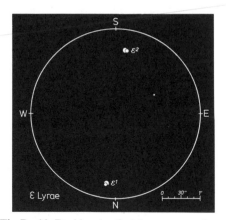

The Double Double, ε (epsilon) Lyrae, as seen through a telescope. *Wil Tirion.*

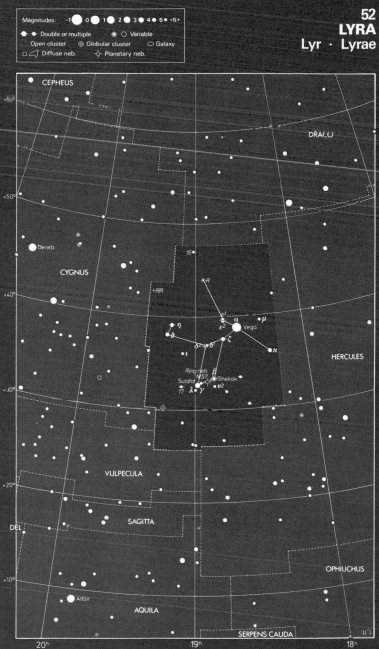

Magnitudes: -1 0 1 2 3 4 5 <5•
•–• Double or multiple   ⊚ ○ Variable
⊙ Open cluster   ⊕ Globular cluster   ◯ Galaxy
▱ Diffuse neb.   ✧ Planetary neb.

CEPHEUS

DRACO

Deneb

CYGNUS

16

R

RR

η
θ
ε¹
ε²
α   Vega
μ
δ² δ
ι   ζ
κ

HERCULES

Ring neb. β
Sulafat M57 Sheliak
17 λ γ ν2

VULPECULA

SAGITTA

DEL

OPHIUCHUS

Altair

AQUILA

SERPENS CAUDA

20ʰ   19ʰ   18ʰ

MENSA   The Table Mountain

A constellation introduced by Lacaille to commemorate the Table Mountain at the Cape of Good Hope, from which he surveyed the southern skies in the 1750s. Part of the Large Magellanic Cloud strays from neighbouring Dorado over the border into Mensa, possibly reminding Lacaille of the cloud that frequently caps the real Table Mountain. Unfortunately, the constellation itself is faint and unimportant.

α (alpha) Mensae, mag. 5.1, is a yellow star similar to the Sun, 28 l.y. away.

β (beta) Men, mag. 5.3, is a yellow star 140 l.y. away.

γ (gamma) Men, mag. 5.2, is an orange giant 420 l.y. away.

η (eta) Men, mag. 5.5, is an orange giant 460 l.y. away.

◄ δ (delta) Lyr is a wide naked eye or binocular double consisting of two unrelated stars: one blue-white, mag. 5.6, 880 l.y. away; and a red giant 720 l.y. away, which varies erratically between mags. 4.5 and 6.5.

ε (epsilon) Lyr, 120 l.y. away, is a celebrated quadruple star commonly called the Double Double. It is easily separated into two stars of mags. 4.7 and 5.1 by binoculars or even keen eyesight. But a telescope of 60–75 mm aperture and high magnification reveals that each star is itself double, with the orientation of the two pairs almost at right angles to each other. Quadruple stars are rare, and this is the finest of them.

ζ (zeta) Lyr, 210 l.y. away, is a double star, easily split in small telescopes or binoculars into components of mags. 4.4 and 5.7.

η (eta) Lyr, 880 l.y. away, is a blue-white star of mag. 4.4 with a wide mag. 8.7 companion visible in small telescopes.

RR Lyr, a giant star on the borders with Cygnus, is the prototype of an important class of variable stars used as 'standard candles' for indicating distances in space. RR Lyrae variables are often found in globular clusters, and are thus known as cluster-type variables. They are giant stars related to Cepheid variables that pulsate in size, varying by about one mag. in less than a day. RR Lyrae itself varies from mags. 7.4–8.6 every 0.57 days.

M 57 (NGC 6720), the Ring Nebula, is a famous planetary nebula 4100 l.y. away, conveniently placed between β (beta) and γ (gamma) Lyrae. On long-exposure photographs with large telescopes it looks like a celestial smoke ring. Small telescopes show it as a noticeably elliptical misty disk, but larger telescopes are needed to see the central hole. It is one of the brightest planetary nebulae, and appears larger in the sky than the planet Jupiter.

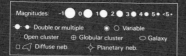

Magnitudes: -1 0 1 2 3 4 5 <5·
Double or multiple ◉ ○ Variable
Open cluster ⊕ Globular cluster ◯ Galaxy
□ ◁ Diffuse neb. ◇ Planetary neb.

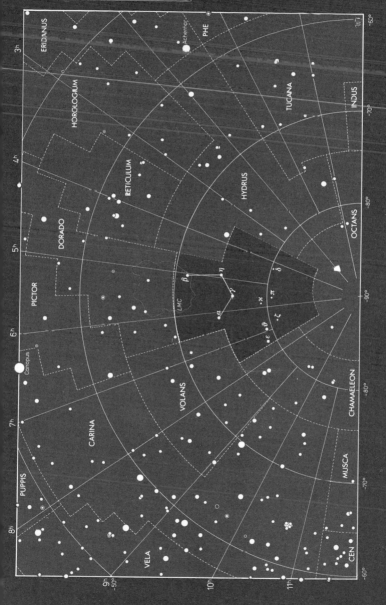

MICROSCOPIUM   The Microscope

Another of the southern hemisphere constellations representing scientific instruments that were introduced in the 1750s by the Frenchman Nicolas Louis de Lacaille. As with so many of his constellations, Microscopium is little more than a filler, encompassing a few faint stars between better-known constellations.

α (alpha) Microscopii, mag. 4.9, is a yellow giant star 240 l.y. away. It has a 10th mag. companion, visible in small telescopes.

γ (gamma) Mic, mag. 4.7, is a yellow giant 230 l.y. away.

ε (epsilon) Mic, mag. 4.7, is a blue-white star 110 l.y. away.

The Rosette Nebula NGC 2237 in Monoceros, perhaps the most strikingly beautiful nebula in the heavens, surrounds the star cluster NGC 2244. For descriptions, see page 180. *Hale Observatories photograph.*

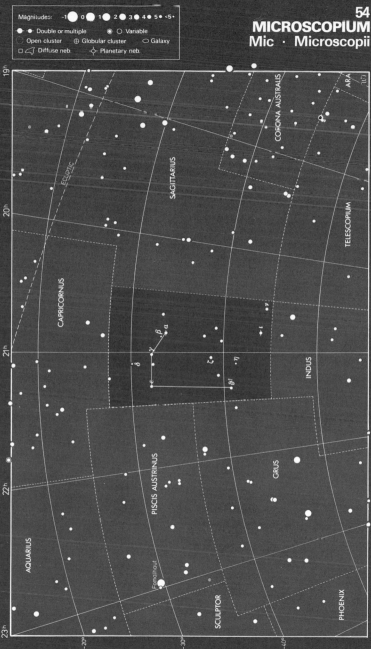

Magnitudes: −1 0 1 2 3 4 5 <5
● Double or multiple   ◉ ○ Variable
○ Open cluster   ⊕ Globular cluster   ○ Galaxy
□ ⌐ Diffuse neb.   ✧ Planetary neb.

ARA

CORONA AUSTRALIS

SAGITTARIUS

ECLIPTIC

TELESCOPIUM

CAPRICORNUS

β α
γ
δ
ε
ζ η
θ ι

INDUS

PISCIS AUSTRINUS

GRUS

AQUARIUS

Fomalhaut

SCULPTOR

PHOENIX

19ʰ
20ʰ
21ʰ
22ʰ
23ʰ

−20°
−30°
−40°

MONOCEROS   The Unicorn

A faint but fascinating constellation between Orion and Canis Minor. Jakob Bartsch, German mathematician and son-in-law of Johannes Kepler, brought it into general use on his star charts of 1624, although there are references to such a constellation in this position in earlier works. Its location in the Milky Way ensures that it is well stocked with nebulae and clusters. Among the most celebrated stars of Monoceros is Plaskett's Star, a 6th magnitude spectroscopic binary named after the Canadian astronomer John S. Plaskett who found in 1922 that it is the most massive pair of stars known; each of their masses are presently estimated as at least 55 times that of the Sun. Plaskett's Star lies at a distance of 3600 l.y. It is near the cluster NGC 2244, of which it may be an outlying member.

α (alpha) Monocerotis, mag. 3.9, is an orange giant star 180 l.y. away.

β (beta) Mon, 720 l.y. away, is rated as perhaps the finest triple star in the sky. The smallest of telescopes should separate the three components, of mags. 4.5, 5.2, and 5.6, on a good night. They form a curving arc of blue-white stars, the faintest two being the closest together.

δ (delta) Mon, mag. 4.2, is a blue-white star 210 l.y. away. It has a wide naked eye companion of mag. 5.5 which is unrelated.

ε (epsilon) Mon, 180 l.y. away, is an easy double star for small telescopes, consisting of yellow and blue components of mags. 4.3 and 6.7 in an attractive low-power field.

S Mon is an intensely luminous blue-white star of mag. 4.7, over 3000 l.y. away, situated in the star cluster NGC 2264 (see below). It is also a double star, with a close companion of mag. 8 visible in small telescopes. To add to the complexity, S Mon is slightly variable, fluctuating erratically by a few tenths of a magnitude.

M 50 (NGC 2323) is a large cluster of about 100 stars, visible in binoculars and small telescopes. Apertures of 100 mm or so resolve it into faint stars with a red star at the centre. M 50 is approximately 2600 l.y. away.

NGC 2232 is a scattered cluster visible in binoculars, centred on the 5th mag. blue-white star 10 Mon, 1000 l.y. away.

NGC 2237–9, NGC 2244 is a complex combination of a faint diffuse nebula, known as the Rosette Nebula, and a cluster of about 16 stars, all about 3600 l.y. away. On long-exposure photographs the nebula appears as a pink loop, twice the apparent diameter of the full Moon. Visual observations with large amateur telescopes reveal only the three brightest parts of the nebula, each of which is given a separate NGC number. The associated cluster, NGC 2244, consists of stars that have been born from the Rosette Nebula's gas; just visible to the naked eye or with binoculars, the cluster of stars is centred on the 6th mag. yellow giant 12 Mon, which in fact seems to be an unrelated foreground star. The cluster is likely to be the only part of this celebrated object visible in small amateur telescopes. (See page 178.)   ▶

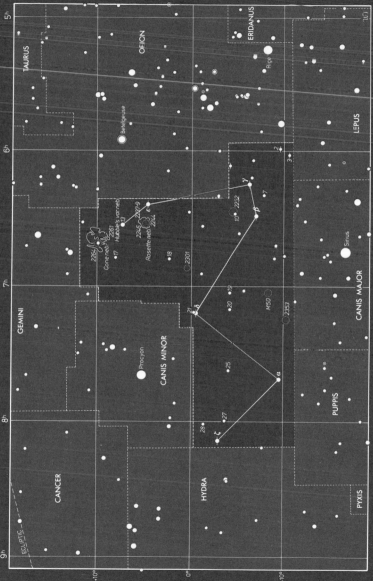

MUSCA   The Fly

A small southern constellation lying at the foot of the Southern Cross. Its origin is attributed to the German astronomer Johann Bayer in 1603, who called it Apis, the Bee, which was subsequently altered to the present title. There is little of note in the constellation other than part of the Coalsack Nebula which spills into Musca from the neighbouring Southern Cross.

α (alpha) Muscae, mag. 2.7, is a blue-white star 330 l.y. away.

β (beta) Mus, 290 l.y. away, is a close pair of stars of mags. 3.9 and 4.2, needing a telescope of 100 mm aperture to split them.

δ (delta) Mus, mag. 3.6, is an orange giant 180 l.y. away.

NGC 4833 is an 8th mag. globular cluster 17,000 l.y. away.

◄ NGC 2261, Hubble's Variable Nebula, is a small, faint nebula containing the remarkable variable star R Mon. At its brightest, this star is of mag. 9.5. Its erratic brightness fluctuations, down to about 12th mag., may be caused by the pangs of its birth from the surrounding nebula. This star and nebula, open to study only by larger amateur telescopes, lie an estimated 6500 l.y. away.

NGC 2264 is another combination of star cluster and nebula. The cluster, visible in binoculars, contains about 20 members including the 5th mag. star S Mon (see above). The nebula, known as the Cone Nebula because of its tapered shape, shows up well only on long-exposure photographs and is beyond the reach of amateur telescopes. The distance of NGC 2264 is about 3000 l.y., almost as far as the Rosette Nebula (NGC 2237).

NGC 2301 is a binocular cluster of 60 or so stars of 8th mag. and fainter, 2500 l.y. away.

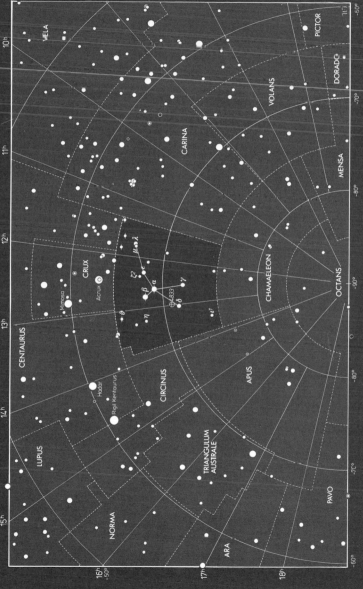

NORMA   The Level

A superfluous constellation invented in the 1750s by Nicolas Louis de Lacaille, who populated the southern skies with several constellations representing scientific instruments, in this case the surveyor's Level. Originally the stars of which it is made were attached to Ara and Lupus. Since Lacaille's time the outline of Norma has been altered, so that the stars which were formerly α (alpha) and β (beta) Normae have now been swallowed by neighbouring constellations. Norma lies in a rich region of the Milky Way.

γ^2 (gamma^2) Normae, mag. 4.0, is a yellow giant star 130 l.y. away. Next to it in the sky lies the far more distant yellow supergiant γ^1 (gamma^1) Normae, mag. 5.0.

δ (delta) Nor, mag. 4.7, is a white star 230 l.y. away.

ε (epsilon) Nor, 490 l.y. away, is a double star with components of mag. 4.8 and 7.5 visible in small telescopes. Each star is also a spectroscopic binary, thus making this a four star system.

ι^1 (iota^1) Nor, 98 l.y. away, appears in small telescopes as a double star of mags. 4.9 and 8.5.

NGC 6087 is a binocular cluster of about 35 stars, 3600 l.y. away.

On the border of Norma with Ara lies an unusual patch of nebulosity called NGC 6164–65. It surrounds a young, hot 7th mag. star known only by its catalogue number of HD 148937. The gas that makes up the nebula has been thrown off in a series of outbursts by the central star. *Anglo–Australian Telescope Board.*

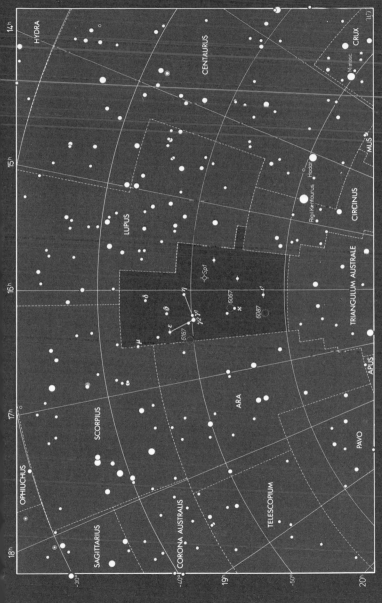

OCTANS   The Octant

The constellation that contains the south pole of the sky. Despite this privileged position, Octans is faint and unremarkable. There is no southern equivalent of Polaris, the North Pole Star; the nearest moderately bright star to the south celestial pole is Sigma Octantis, mag. 5.5, which lies about 1° from the pole. The constellation of Octans commemorates the instrument known as the octant, a forerunner of the sextant, invented by the Englishman John Hadley and used by him for measuring star positions. The constellation itself was introduced in the 1750s by Nicolas Louis de Lacaille during his stay at the Cape of Good Hope, and its dullness is a memorial to his dreadful lack of imagination.

α (alpha) Octantis, mag. 5.2, is a white star 230 l.y. away.

β (beta) Oct, mag. 4.2, is a white star 65 l.y. away.

δ (delta) Oct, mag. 4.3, is an orange giant 200 l.y. away.

θ (theta) Oct, mag. 4.8, is an orange giant 250 l.y. away.

λ (lambda) Oct, 36 l.y. away, is a double star with components of mags. 5.5 and 7.6, individually visible in small telescopes.

ν (nu) Oct, mag. 3.8, is an orange giant 105 l.y. away.

Detail chart showing the position of the south celestial pole and its movement over a period of a century as a result of precession. *Wil Tirion.*

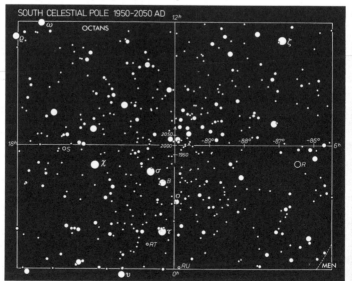

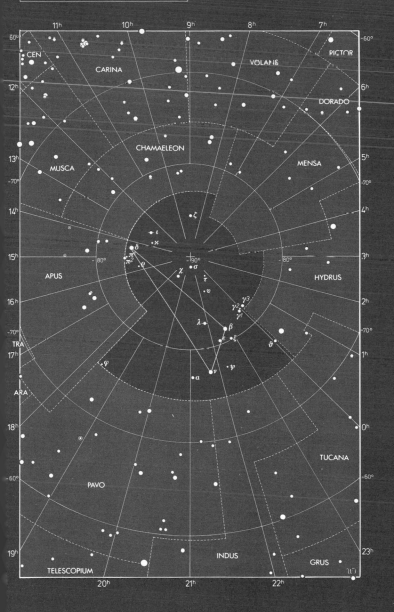

OPHIUCHUS   The Serpent Holder

An ancient constellation, representing a man encoiled by a serpent (the constellation Serpens). Ophiuchus is identified by some as Aesculapius, a mythological healer and forerunner of Hippocrates. Aesculapius served as ship's doctor on the Argo. Another interpretation sees this constellation as Enkidu, the companion of Gilgamesh, who circles head-to-head with Ophiuchus in the sky, in the form of the constellation Hercules. Perhaps the most celebrated star in Ophiuchus is a 9th magnitude red dwarf too faint to be seen by the naked eye: Barnard's Star, at 6 l.y. away the 2nd-closest star to the Sun. It is named after the American astronomer Edward Emerson Barnard, who found in 1916 that it has the greatest proper motion of any star; it covers a distance equal to the apparent diameter of the Moon every 180 years. Barnard's Star is of particular interest because it is believed to have planets. The southern regions of Ophiuchus extend into rich star fields of the Milky Way, looking towards the centre of the Galaxy; consequently the constellation is noted for its star clusters. Ophiuchus was the site of the last supernova seen to erupt in our Galaxy, which appeared between the star ζ (xi) Ophiuchi and the border with Sagittarius in 1604.

α (alpha) Ophiuchi (Ras Alhague, head of the serpent charmer), mag. 2.1, is a white star 62 l.y. away.

β (beta) Oph, mag. 2.8, is a yellow giant 120 l.y. away.

γ (gamma) Oph, mag. 3.8, is a blue-white star 115 l.y. away.

δ (delta) Oph (Yed Prior, the preceding star of the hand), mag. 2.7, is an orange giant 140 l.y. away.

ε (epsilon) Oph (Yed Posterior, the following star of the hand), mag. 3.2, is an orange giant 105 l.y. away.

ζ (zeta) Oph, mag. 2.6, is a blue-white star 550 l.y. away.

η (eta) Oph, mag. 2.4, is a blue-white star 59 l.y. away.

ϱ (rho) Oph, 750 l.y. away, is a striking multiple star for small telescopes. The brightest star is mag. 5.2. This has a close companion of mag. 5.9; either side of this close pair are wide companions of mag. 7.1 and 6.6.

τ (tau) Oph, 72 l.y. away, is a close pair of cream-coloured stars, mags. 5.3 and 6.0, orbiting every 280 years and currently closing together, so that they require at least 75 mm aperture.

36 Oph, 18 l.y. away, is a pair of mag. 5.3 orange dwarf stars comfortably split by small apertures.

70 Oph, 17 l.y. away, is a celebrated double star, consisting of yellow and orange components of mags. 4.3 and 6.0 which orbit each other every 88 years. A tele- ▶

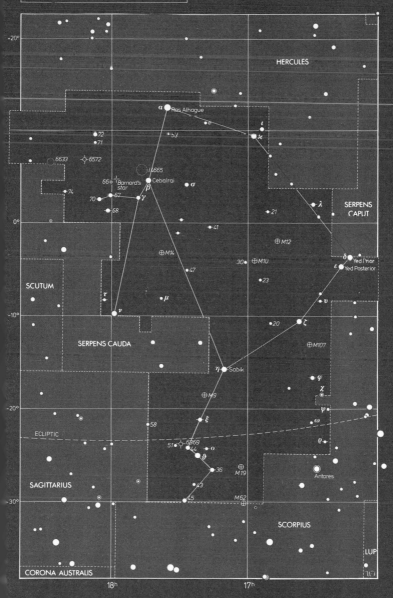

Magnitudes: -1 0 1 2 3 4 5 6

● Double or multiple   ◉ Variable
◌ Open cluster   ⊕ Globular cluster   ◯ Galaxy
□ Diffuse neb.   ✧ Planetary neb.

+20°

HERCULES

α Ras Alhague

ι

κ

+53

72
71

6633   ✧ 6572

IC4665

66   Barnard's star

Cebalrai

β

σ

74

λ

SERPENS CAPUT

70   67

68

γ

21

0°

41

⊕ M12

⊕ M14

30   ⊕ M10

δ Yed Prior

ε Yed Posterior

47

23

υ

τ

μ

SCUTUM

ν

-10°

20   ζ

SERPENS CAUDA

⊕ M107

η Sabik

φ

χ

⊕ M9

ψ

-20°

ξ

ω

58

ECLIPTIC

51   6369

44   o

ϑ

36

⊕ M19

SAGITTARIUS

43

Antares

45   M62 ⊕

-30°

SCORPIUS

LUP

CORONA AUSTRALIS

18ʰ

17ʰ

ORION   The Hunter

Without doubt the brightest and grandest constellation in the sky, crammed with objects of interest for all sizes of instrument. Orion's impressiveness stems in large measure from the fact that it is an area of star formation in a nearby arm of the Galaxy, centred on the famous Orion Nebula. The Orion Nebula, M 42, marks the Hunter's sword, hanging from his belt. The belt itself is formed by a line of three bright stars. Orion is depicted as brandishing a club and shield at the snorting Taurus the Bull. In one story the boastful Orion was stung to death by a scorpion, and is now placed in the sky so that he sets as the constellation Scorpius rises. Each year the Orionid meteors radiate from a point near the border with Gemini. As many as 20 Orionid meteors per hour may be seen around October 21.

α (alpha) Orionis (Betelgeuse), 310 l.y. away, is a red supergiant star so large that it is unstable. It fluctuates erratically in size from about $300-400\times$ the Sun's diameter, changing in brightness as it does so from mags. 0.4 to 1.3.

β (beta) Ori (Rigel, giant's leg), at mag. 0.1 the brightest star in Orion, is a blue-white supergiant 910 l.y. away; note its colour contrast with Betelgeuse. Rigel has a 7th mag. companion, difficult to see in small telescopes, particularly in poor seeing, because of glare from the primary.

γ (gamma) Ori (Bellatrix, the conqueror), mag. 1.6, is a blue giant star 360 l.y. away. ▶

◀ scope of 100 mm aperture and high magnification is needed to split them. This system was once suspected of having planets, but this is no longer thought to be the case.

M 10 (NGC 6254) is a 7th mag. globular cluster visible in binoculars or a small telescope. It is approximately 19,000 l.y. away, a little farther than its neighbour M 12. Individual stars can be resolved with telescopes of 150 mm aperture or more.

M 12 (NGC 6218) is a 7th mag. globular cluster visible in binoculars or a small telescope. It appears slightly larger than its neighbour M 10, and its stars are more loosely scattered. There are several other globular clusters in Ophiuchus worthy of attention, but M 10 and M 12 are the finest.

NGC 6572 is a 10th mag. planetary nebula, 2500 l.y. away, visible in at least 75 mm aperture as a tiny blue-green ellipse.

NGC 6633 is a scattered cluster of about 65 stars of mag. 7 and fainter, visible in binoculars, 1600 l.y. away.

IC 4665 is a loose and irregular cluster of a dozen or so stars of mag. 7 and fainter, 1000 l.y. away, best seen in binoculars.

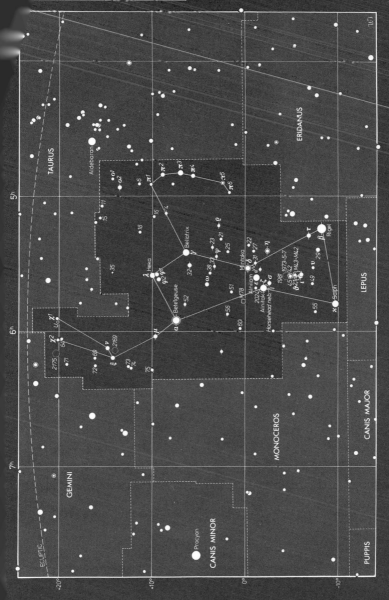

◀ δ (delta) Ori (Mintaka, the belt), 2300 l.y. away, is a complex multiple st the naked eye it appears as a blue–white star of mag. 2.2. Binoculars or smal scopes reveal a wide companion of mag. 6.9. In fact, the main star is als eclipsing binary that varies by about 0.1 of a mag. every 5.7 days.

ε (epsilon) Ori (Alnilam, string of pearls), mag. 1.7, is a blue supergiant, 1200 . away.

ζ (zeta) Ori (Alnitak, the girdle), 1100 l.y. away, appears to the naked eye as blue–white star of mag. 2.0, but telescopes of 75 mm aperture and above reveal that it has a close companion of mag. 4.2. There is also a much wider 10th mag. star.

η (eta) Ori, 750 l.y. away, is another complex multiple-variable. Telescopes of at least 100 mm aperture at high magnification are needed to show that it consists of two close stars, of mags. 3.7 and 5.1. The brighter star is also an eclipsing binary, varying by about 0.1 of a mag. every 8 days.

θ^1 (theta^1) Ori, 1300 l.y. away, is a multiple star at the heart of the Orion Nebula, from which it has recently formed and which it illuminates. This star is commonly known as the Trapezium, because small telescopes show four stars here; but telescopes of 100 mm aperture also reveal two others of 11th mag. The four main stars of the Trapezium are of mags. 5.1, 6.7, 6.7, and 8.0. Nearby lies θ^2 (theta^2) Orionis, a wide double of mags. 5.1 and 6.4.

ι (iota) Ori, 1800 l.y. away, is a double star on the southern edge of the Orion Nebula, divisible in small telescopes. Its components are of mags. 2.9 and 7.4. Also visible in the same field is a wider double of blue–white stars, Struve 747, of mags. 4.7 and 5.6.

κ (kappa) Ori (Saiph, sword), mag. 2.1, is a blue supergiant, 1300 l.y. away.

λ (lambda) Ori, 1400 l.y. away, is a blue–white star of mag. 3.7 with a close mag. 5.6 companion visible in small telescopes at high magnification.

ϱ (rho) Ori, 280 l.y. away, is an orange giant of mag. 4.5 with a mag. 8.6 companion; the brightness difference makes this pair difficult in smallest telescopes.

σ (sigma) Ori, 1800 l.y. away, is perhaps the most impressive and colourful of all Orion's stellar treasures. To the naked eye it appears as a blue–white star of mag. 3.7, but small telescopes reveal much more. The star has bluish and reddish companions of mags. 6.5 and 7.2 respectively, the brighter of which can be glimpsed in binoculars; closer still is a 10th mag. companion, which is more difficult to see because of glare from the primary. To complete the picture, in the same telescopic field of view is a faint triple star called Struve 761, consisting of a triangle of 8th and 9th mag. stars. Altogether an extraordinarily rich and unexpected sight, to be returned to again and again.

M 42, M 43 (NGC 1976, NGC 1982) is the most celebrated of Orion's deep-sky wonders – a gigantic nebula of gas and dust, 1300 l.y. away and 15 l.y. in diameter, from which a star cluster is being born. Behind the visible part of the nebula, illuminated by the stars of the Trapezium (see θ (theta) Orionis), radio astronomers have detected an even larger dark cloud in which more stars are forming. M 42 covers an area of more than 1° × 1°, and is indisputably the finest diffuse nebula in the sky. Clearly visible to the naked eye as a hazy cloud, binoculars and

...scopes reveal some of the more prominent wreaths and swirls of gas,
...ecome more complex and more breathtaking the larger the aperture.
...gh colour photographs depict the nebula as red and blue, to the eye
...ears distinctly greenish because of the different colour sensitivities of
...graphic film and the human eye. A dark lane of gas known as the Fish
...th separates M 42 from M 43, a smaller and rounder patch that is really a
...ε of the same cloud; M 43 is centred on a 9th mag. star.

...78 (NGC 2068) is a small, elongated nebula centred on a 9th mag. double star.
...n large amateur telescopes it shows wispy structure.

NGC 1977 is an elongated nebulosity just north of the Orion Nebula, centred on
the 5th mag. star 42 Orionis, also known as c Orionis. This object would be
more celebrated if it were not so overshadowed by M 42.

NGC 1981 is a scattered cluster to the north of the nebulosity NGC 1977.
Included in this cluster is the double star Struve 750, a neat pair of 6th and 8th
mag. stars.

NGC 2024 is a glowing area of gas about 0.5° wide, surrounding the star ζ (zeta)
Orionis. Running south from ζ (zeta) Ori is a strip of nebulosity into which is
indented the celebrated Horsehead Nebula, a dark cloud of obscuring dust
shaped like a horse's head. Although long-exposure photographs show NGC
2024 and the Horsehead Nebula well, they are notoriously difficult to see with
amateur telescopes.

Detailed chart of the Orion Nebula region. *Wil Tirion.*

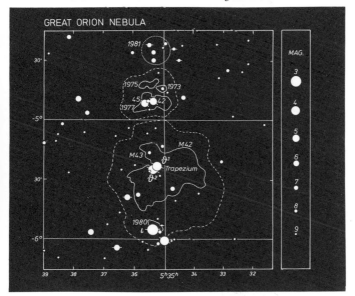

PAVO   The Peacock

A constellation introduced on the 1603 star map of the German ce. cartographer Johann Bayer. It is one of several celestial birds in region, including Apus, Tucana, Grus, and Phoenix. In mythole the peacock was sacred to Juno, goddess of the heavens from whe breast the Milky Way sprang. According to legend, Juno set a creatu with a hundred eyes called Argus to watch over a white heifer; Jun guessed that this heifer was the form into which her husband Jupiter had turned one of his illicit lovers, the nymph Io. At Jupiter's request, Mercury decapitated the watchful Argus and released the heifer. Juno placed the 100 eyes of Argus on the peacock's tail.

α (alpha) Pavonis (Peacock), mag. 1.9, is a blue-white star 230 l.y. away.

β (beta) Pav, mag. 3.4, is a white star 91 l.y. away.

δ (delta) Pav, mag. 3.6, is a yellow star 19 l.y. away.

η (eta) Pav, mag. 3.6, is an orange giant 150 l.y. away.

κ (kappa) Pav, 650 l.y. away, is one of the brightest Cepheid-type variables in the sky. It is a yellow supergiant, varying between mags. 3.9–4.8 every 9.1 days.

ξ (xi) Pav, 310 l.y. away, is a red giant of mag. 4.4 with a close companion of mag. 8.6, lost in the primary's glare in small telescopes.

NGC 6752 is a large globular cluster, visible in binoculars, covering nearly half the apparent diameter of the Moon. It lies over 20,000 l.y. away.

NGC 6744 in Pavo, a stately 11th mag. galaxy with a short central bar and extensive spiral arms. *Anglo-Australian Telescope Board.*

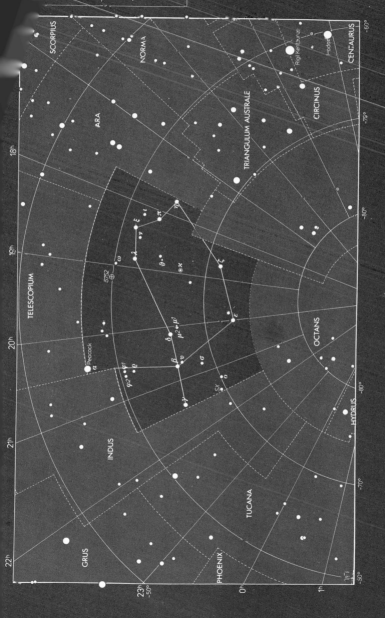

Double or multiple ⊙ ○ Variable
cluster ⊕ Globular cluster ○ Galaxy
Diffuse neb. ◇ Planetary neb.

SCORPIUS

NORMA

ARA

TRIANGULUM AUSTRALE

CENTAURUS

Rigil Kentaurus

CIRCINUS

TELESCOPIUM

5752

ξ ι
π η
ν ξ
ω λ
θ χ
ζ
ε

δ
μ² μ¹
ν
β σ
κ
γ
Sk δ

α
Peacock
φ² φ¹
ρ

OCTANS

INDUS

HYDRUS

TUCANA

GRUS

PHOENIX

18h
19h
20h
21h
22h
23h
0h
1h

-60°
-70°
-80°
-50°

PEGASUS

The winged horse of Greek mythology, born from the blood of Me
after she was slain by Perseus, who lies nearby in the sky. The r.
famous feature of Pegasus is its great square, outlined by four st
One of these stars is now assigned to Andromeda, although astronome
of old knew it as δ (delta) Pegasi. The great square of Pegasus contain
surprisingly few naked-eye stars for such a large area of sky. In fact,
the entire constellation is poor in objects of interest, apart from a major
globular cluster, M 15.

α (alpha) Pegasi (Markab, meaning saddle), mag. 2.5, is a blue-white star 100 l.y.
away.

β (beta) Peg (Scheat, shoulder), 180 l.y. away, is a red giant 90× the Sun's
diameter, which varies from mag. 2.4 to mag. 2.8, about every month.

γ (gamma) Peg (Algenib, the wing or side), mag. 2.8, is a blue-white star 490 l.y.
away. It is a variable of the β (beta) Cephei type but its light variations every
3 hrs. 40 mins are too small to be noticeable to the naked eye.

ε (epsilon) Peg (Enif, the nose), 520 l.y. away, is a yellow supergiant of mag.
2.4. Small telescopes, or even good binoculars, reveal a wide bluish mag. 8.7,
companion star. Larger telescopes also show an 11th mag. companion closer to
ε (epsilon) Pegasi, making this an apparent triple system.

ζ (zeta) Peg (Homam), mag. 3.4, is a white star 160 l.y. away.

η (eta) Peg (Matar), mag. 2.9, is a yellow giant 170 l.y. away.

π (pi) Peg is a very wide binocular duo of white and yellow stars. They are of
mags. 4.3 and 5.6, and lie 310 and 320 l.y. away respectively.

1 Peg, mag. 4.1, is a yellow giant 210 l.y. away, with a 9th mag. companion visible
in small telescopes.

M 15 (NGC 7078) is an outstanding, bright, globular cluster 50,000 l.y. distant,
at the limit of naked eye visibility but easily seen in binoculars, with a nearby
6th mag. star as a sure guide to its location. Telescopes show it as a glorious misty
sight in an attractive field. With apertures of 150 mm or so its outer regions can
be resolved into a mottled ground of sparkling stars, and larger apertures show
stars all the way to the bright and condensed core.

NGC 7331 is a 10th mag. spiral galaxy, visible under good conditions as an
elongated smudge in apertures of 100 mm or so. It lies about 40 million l.y. away.

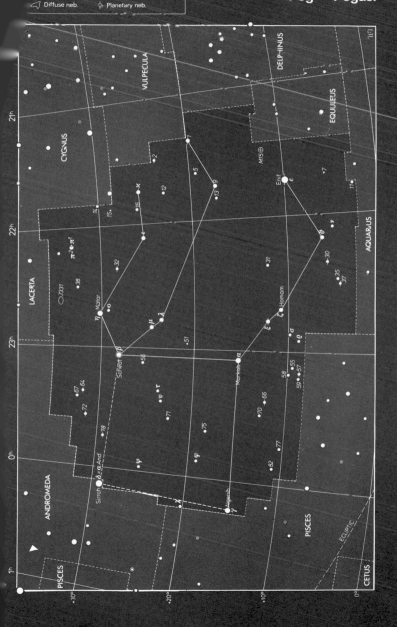

PERSEUS

Perseus was the hero of Greek mythology who rescued the chai
Andromeda from the clutches of the sea monster, Cetus. Previou
Perseus had slain Medusa the Gorgon, whose head he is picture
holding in one hand. The Gorgon's evil eye is represented by the wink
ing star Algol. Perseus lies in a rich part of the Milky Way, and is wel
worth sweeping with binoculars. In 1901 a brilliant nova flared up to
magnitude 0.2 between the stars δ (delta) and β (beta) Persei; this nova
threw off a shell of gas that is now visible in large telescopes. Near γ
(gamma) Persei lies the radiant of the Perseid meteors, the most glorious
meteor shower of the year: around August 12–13 as many as 60 bright
meteors per hour can be seen streaming from Perseus. Between α (alpha)
and β (beta) Persei lies the radio source Perseus A, associated with the
13th magnitude supergiant elliptical galaxy NGC 1275, which lies at
the centre of the Perseus cluster of galaxies, 300 million l.y. away.

α (alpha) Persei (Algenib, meaning side, or Mirfak, the elbow), mag. 1.8, is a
yellow supergiant 620 l.y. away. Binoculars reveal a brilliant scattering of stars
in this region.

β (beta) Per (Algol, the demon), 95 l.y. away, is one of the most celebrated variable
stars in the sky. It is the prototype of the eclipsing binary class of variables, in
which two close stars periodically eclipse one another as they orbit their common
centre of gravity. In the case of Algol, the eclipses occur every 2.87 days, when
the star's apparent brightness sinks from mag. 2.2 to 3.5 for a period of 10 hours
before returning to maximum.

γ (gamma) Per, mag. 2.9, is a yellow giant 110 l.y. away.

δ (delta) Per, mag. 3.0, is a blue giant 330 l.y. away.

ε (epsilon) Per, 680 l.y. away, is a blue–white star of mag. 2.9 with a mag. 8.1 companion, difficult to see in the smallest telescopes because of the magnitude
contrast.

ζ (zeta) Per, 1100 l.y. away, is a blue supergiant of mag. 2.9 with a mag. 9.4
companion visible in small telescopes.

η (eta) Per, 820 l.y. away, is an orange supergiant of mag. 3.8 with a mag. 8.6 blue
companion that forms an attractive double for small telescopes. The same field
of view contains a sprinkling of background stars.

ϱ (rho) Per, 200 l.y. away, is a red giant that varies between mags. 3.3 and 4.0 in
semi–regular fashion.

NGC 869, NGC 884, the famous Double Cluster in Perseus, also known as
h and χ (chi) Persei. They are two open star clusters visible to the naked eye and
superb in binoculars, each covering an area larger than the full Moon. NGC 869
is the brighter and richer of the pair, containing an estimated 350 stars compared
with its companion's 300 stars. They both lie approx. 7300 l.y. away. Small
telescopes have an advantage when observing these objects, for with low powers

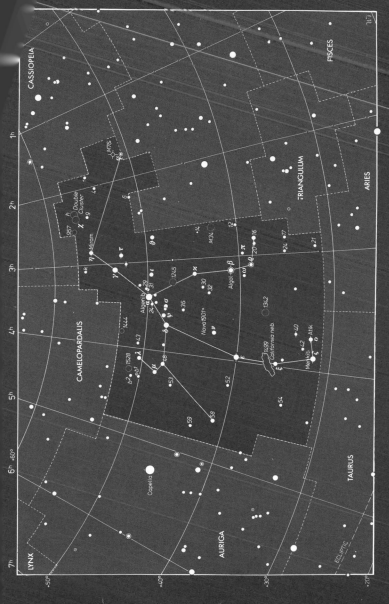

es: -1 0 1 2 3 4 5 <5

ouble or multiple Variable
en cluster Globular cluster Galaxy
Diffuse neb. Planetary neb.

CASSIOPEIA

PISCES

TRIANGULUM

ARIES

M76

h
Double
Cluster
χ
957
4

η Mirfak
τ
γ
29
34 α Algenib
31
κ
σ
ψ
δ
1444
43
1528
b²
b¹
μ
48
53
λ

ϑ

6

14
M34
12

π
20 16
17
ι ο
ω
24 21

Algol β

κ
1245
30
32
ψ
1342

Nova 1901ᵃ

1499
California neb.
42
ξ
ε
40
Atik
ζ ο
Menkib

CAMELOPARDALIS

36

52

54

58
59

Capella

TAURUS

AURIGA

LYNX

ECLIPTIC

1ʰ
2ʰ
3ʰ
4ʰ
5ʰ
6ʰ -60°
7ʰ

-50°
-40°
-30°
-20°

PHOENIX   The Phoenix

An inconspicuous constellation lying near the southern end of Erida[...]
representing the mythological bird that was regularly reborn from[...]
own ashes. It was introduced in 1603 by Johann Bayer in an area t[...]
had been known by the Arabs as the Boat, moored on the shores of th[...]
river; later this figure was seen as an eagle or a griffin, so the associatio[...]
of this area with a bird was well established by Bayer's time.

α (alpha) Phoenicis, mag. 2.4, is a yellow giant star 78 l.y. away.

β (beta) Phe, 130 l.y. away, appears to the naked eye as a yellow star of mag. 3.3.
In fact, it is a close double with well-matched components each of mag. 4.1,
divisible in telescopes of 100 mm aperture.

γ (gamma) Phe, mag. 3.4, is a red supergiant 910 l.y. away.

ζ (zeta) Phe, 220 l.y. away, is a complex variable and multiple star. The main star
is a blue-white eclipsing variable which fluctuates around 4th mag. every 1.67
days. It has a mag. 8.2 companion visible in small telescopes. There is also a
much closer 7th mag. companion which requires telescopes above 200 mm
aperture.

◄ both clusters can be seen in the same field of view, which is not always the case
with larger and more powerful telescopes. Most of the stars in the cluster are
blue-white, but there are several red stars to be spotted among the masses of
glittering points resolved by telescopes. A breathtaking sight in all apertures.

M 34 (NGC 1039) is a bright star cluster at the limit of naked eye visibility. It is
far less rich and condensed than the Double Cluster, containing about 80 stars
splashed over an area similar to the apparent size of the full Moon. Binoculars just
resolve it into stars, and it is well seen in small telescopes. M 34 lies about 1400
l.y. away.

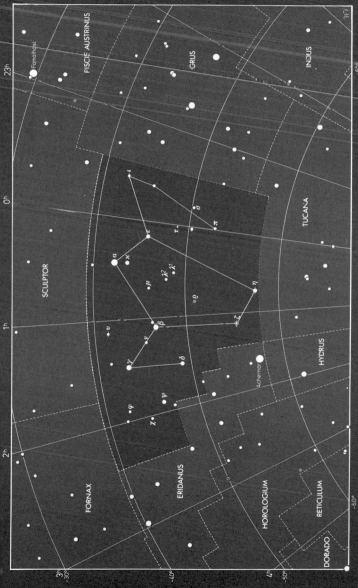

PICTOR The Painter's Easel

A faint constellation overshadowed by the neighbouring brilliant star Canopus in Carina to one side, and the Large Magellanic Cloud in Dorado on the other. The constellation was invented in the 1750s by Nicolas Louis de Lacaille, who originally called it Equuleus Pictoris, which has since been shortened to the current name. It contains the 9th magnitude red dwarf known as Kapteyn's Star, 12.7 l.y. away, named after the Dutch astronomer who discovered in 1897 that it has the 2nd-largest proper motion of any known star (the record is held by Barnard's Star in Ophiuchus).

α (alpha) Pictoris, mag. 3.3, is a white star 52 l.y. away.

β (beta) Pic, mag. 3.9, is a white star 78 l.y. away.

γ (gamma) Pic, mag. 4.5, is an orange giant 260 l.y. away.

δ (delta) Pic, mag. 4.8, is a blue-white star of uncertain distance.

M 74 in Pisces is a spiral galaxy with loosely wound arms. For a description, see page 204. *Hale Observatories photograph.*

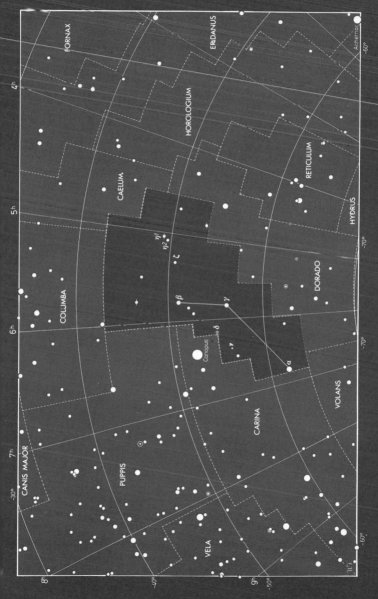

PISCES The Fishes

An ancient constellation representing a pair of fishes tied by their tails, the knot being marked by the star α (alpha) Piscium. One legend identifies the constellation with Venus and her son Cupid, who turned themselves into fishes and swam away from the attack of the monster Typhon. The constellation's most celebrated feature is that it contains the vernal equinox – the point at which the Sun moves across the celestial equator into the northern celestial hemisphere each year. This point originally lay in Aries, but has now moved into Pisces because of precession; eventually it will move on into Aquarius. Though faint, Pisces contains numerous stars of interest.

α (alpha) Piscium (Al Rischa, the cord), 98 l.y. away, is a challenging double star with an orbital period of 720 years. At present, the components are closing and need at least a 75 mm telescope and high magnification to separate them. The stars, of mags. 4.3 and 5.2, are each spectroscopic binaries. Their colour is bluish-white, although some observers see the brighter star as greenish.

β (beta) Psc, mag. 4.5, is a blue-white star 320 l.y. away.

γ (gamma) Psc, mag. 3.7, is a yellow giant 160 l.y. away.

ζ (zeta) Psc, 110 l.y. away, is a wide double of mag. 4.9 and 6.3 stars, visible in the smallest telescopes.

η (eta) Psc, mag. 3.6, is the brightest star in the constellation. It is a yellow giant, 140 l.y. away.

κ (kappa) Psc, mag. 4.9, is a blue-white star 98 l.y. away, with a 6th mag. binocular companion, unrelated.

ϱ (rho) Psc, mag. 5.4, is a white star 98 l.y. away which forms an easy binocular duo with the unrelated orange giant star 94 Piscium, mag. 5.5, 390 l.y. away.

ψ¹ (psi¹) Psc, 390 l.y. away, is a wide double consisting of blue-white stars of mags. 5.3 and 5.6, visible in small telescopes or even good binoculars.

19 Psc (TX Psc) is a deep red irregular variable star visible in binoculars, which fluctuates around 5th or 6th mag.

M 74 (NGC 628) is a face-on spiral galaxy 22.5 million l.y. away. At 10th mag. it is one of the faintest objects on Messier's list. Although it can be glimpsed in small instruments under dark conditions, it needs at least a 150 mm telescope to be well seen. (See page 202.)

Magnitudes: -1 ● 0 ● 1 ● 2 ● 3 ● 4 ● 5 • <5 ·
● ● Double or multiple   ◉ ○ Variable
○ Open cluster   ⊕ Globular cluster   ○ Galaxy
□ ◁ Diffuse neb.   ◇ Planetary neb.

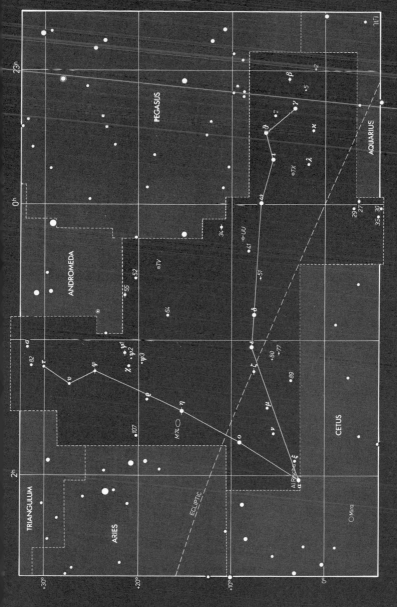

PISCIS AUSTRINUS   The Southern Fish

Sometimes also called Piscis Australis, this constellation has been known since ancient times and is often represented as a fish drinking the flow of water from the urn of neighbouring Aquarius. This fish has been identified with the Babylonian fish god Oannes, and is said to be the parent of the zodiacal fish, Pisces.

α (alpha) Piscis Austrini (Fomalhaut, fish's mouth), mag. 1.2, is a blue-white star 22 l.y. away.

β (beta) PsA, 170 l.y. away, is a wide double star consisting of a mag. 4.3 white primary and an optical (unrelated) mag. 7.8 companion, visible in small telescopes.

γ (gamma) PsA, 190 l.y. away, is a double star of mags. 4.5 and 8.1, made difficult to split in small telescopes by the magnitude contrast.

η (eta) PsA, 420 l.y. away, is a close pair of blue-white stars of mags. 5.8 and 6.8, divisible with an aperture of 100 mm and high power.

Although Piscis Austrinus contains no notable deep-sky objects, its neighbour Sculptor has several galaxies of note, among them NGC 253, shown here, a spiral galaxy seen nearly edge-on. It has no central bulge but its arms are a seething mass of stars and dust. See also page 224. *Anglo-Australian Telescope Board.*

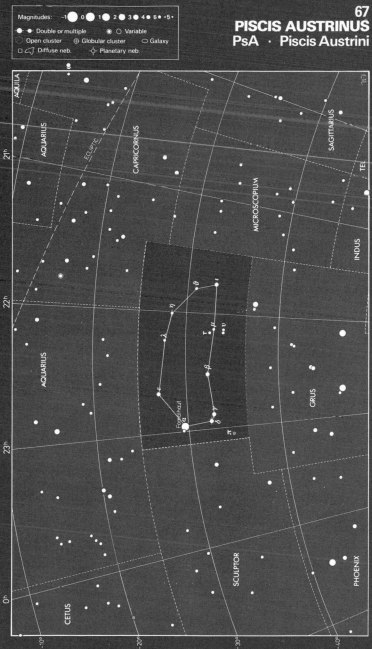

PUPPIS   The Stern

This is the largest of the four sections into which the ancient con-stellation of Argo Navis, the ship of the Argonauts, was dismembered in the 1750s by Nicolas Louis de Lacaille; the other sections are Carina, Pyxis, and Vela. Puppis lies in the Milky Way and contains rich star-fields for sweeping with binoculars.

ζ (zeta) Puppis (Naos, ship), mag. 2.3, is a brilliant blue-white star 1500 l.y. away, one of the most intensely hot stars known, with a surface temperature of about 35,000°C.

ξ (xi) Pup, mag. 3.3, is a yellow supergiant 750 l.y. away. Binoculars reveal a wide, unrelated mag. 5.3 orange companion, 320 l.y. away.

π (pi) Pup, mag. 2.7, is an orange giant 130 l.y. away.

ϱ (rho) Pup, mag. 2.8, is a yellow giant 300 l.y. away. It is a variable of the δ (delta) Scuti type, changing in brightness by 0.1 of a mag. every 3 hrs. 23 mins.

τ (tau) Pup, mag. 2.9, is a yellow giant 82 l.y. away.

k Pup, 360 l.y. away, is a striking double star with near-identical blue-white components of mags. 4.5 and 4.6 easily divisible in small telescopes.

L Pup is an optical double, consisting of two unrelated stars. L^1 is mag. 4.9, a blue-white star 550 l.y. away. L^2 is a red giant semi-regular variable, 180 l.y. away, that fluctuates between 3rd and 6th mag. about every 140 days.

V Pup, 1500 l.y. away, is an eclipsing binary of the β (beta) Lyrae type. It varies from mags. 4.5–5.1 every 35 hours.

M 46 (NGC 2437) is an 8th mag. cluster of about 150 faint stars of remarkably uniform brightness, most being around 10th mag. Small telescopes show M 46 as a sprinkling of stardust. The cluster lies about 6000 l.y. away. On the north edge of the cluster lies the 11th mag. planetary nebula NGC 2438. This is not associated with the cluster, but is a foreground object, about 3000 l.y. away.

M 47 (NGC 2422) is a naked eye cluster of about 50 stars, the brightest being of 6th mag. The cluster lies 3800 l.y. away.

M 93 (NGC 2447) is a binocular cluster, 3600 l.y. away, consisting of 60 stars of mag. 8 and fainter, arranged in a wedge shape.

NGC 2477 is a large binocular cluster of about 300 faint stars, like a loose globular, 6200 l.y. away.

NGC 2451 is a large and bright cluster of 50 stars centred on the mag. 3.6 orange supergiant star c Puppis, 1700 l.y. away.

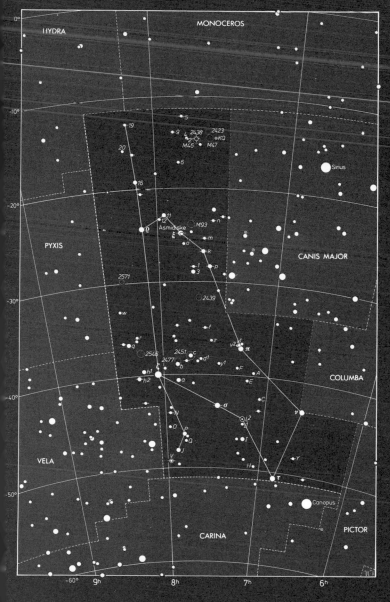

Magnitudes: −1 ● 0 ● 1 ● 2 ● 3 ● 4 ● 5 ● <5 ·

●–● Double or multiple   ◉ ○ Variable
◎ Open cluster   ⊕ Globular cluster   ◯ Galaxy
□ ⊂⊐ Diffuse neb.   ◇ Planetary neb.

HYDRA

MONOCEROS

0°

−10°

HYDRA

5
19
9   2438 2423
M46 M47 KQ
20
6

Sirius

16

−20°

PYXIS

11
12
b Asmidiske
ξ   M93 n
o   m
k
1 p
3

CANIS MAJOR

2571

−30°

2439

w

q

f
7
2545 c   v2 π
2451 d1
2477 b y1
h1   F A
h2 a   E

COLUMBA

−40°

σ C
N   L2 v
O   M
P I
Q Y
J H
V τ

VELA

Canopus

−50°

PICTOR

CARINA

−60°

9ʰ   8ʰ   7ʰ   6ʰ

PYXIS   The Compass

The smallest and least impressive of the four parts into which the ancient constellation of Argo Navis, the Ship, was divided in the 1750s by Nicolas Louis de Lacaille; the other sections are Carina, Puppis, and Vela. There are no objects of particular interest to users of small telescopes in this faint constellation, despite the fact that it lies in the Milky Way.

α (alpha) Pyxidis, mag. 3.7, is a blue-white star 1300 l.y. away.

β (beta) Pyx, mag. 4.0, is a yellow giant 150 l.y. away.

γ (gamma) Pyx, mag. 4.0, is an orange giant 240 l.y. away.

T Pyx is a recurrent nova that has erupted more often than any other nova, in 1890, 1902, 1920, 1944 and 1966. Normally it is of mag. 14, but brightens to 6th–8th mag. It lies between β (beta) and λ (lambda) Pyx. Further outbursts may be expected.

Nicolas Louis de Lacaille (1713–1762)

Lacaille, a French astronomer, was the first person to map the southern skies comprehensively, as a result of which he became known as the Father of Southern Astronomy. He directed an expedition of the French Academy of Sciences to the Cape of Good Hope in 1750–54, where he systematically surveyed the southern celestial hemisphere, listing over 10,000 stars. The accurate positions of 2000 of these, along with a star map, were published posthumously in 1763 under the title *Coelum Australe Stelliferum*. Lacaille is usually best remembered for the 14 new constellations he introduced, representing instruments used in science and the fine arts: Antlia, Caelum, Circinus, Fornax, Horologium, Mensa, Microscopium, Norma, Octans, Pictor, Pyxis, Reticulum, Sculptor and Telescopium. Lacaille also dismantled one constellation, Robur Carolinum, Charles's Oak; this was formed in 1679 by Edmond Halley from the stars of Argo Navis to commemorate the oak tree in which his patron, King Charles II, hid after his defeat by Oliver Cromwell at the battle of Worcester. It is said that this inspired piece of flattery earned Halley his master's degree from Oxford by the King's express command. Lacaille, less impressed, uprooted the oak and returned its stars to Argo Navis.

Magnitudes: −1 0 1 2 3 4 5 <5

● ● Double or multiple   ◉ ○ Variable
○ Open cluster   ⊕ Globular cluster   ○ Galaxy
▭ ⌐ Diffuse neb.   ✧ Planetary neb.

RETICULUM The Net

A constellation introduced in the 1750s by Lacaille to commemorate an instrument known as the reticle which he used for measuring star positions on his surveys of the southern sky. It is not a prominent constellation, but lies near the Large Magellanic Cloud.

α (alpha) Reticuli, mag. 3.4, is a yellow giant star 390 l.y. away.

β (beta) Ret, mag. 3.9, is an orange star 55 l.y. away.

ζ (zeta) Ret, 40 l.y. away, is a wide naked eye or binocular double of identical yellow stars similar to the Sun, of mags. 5.2 and 5.5.

Over the borders of Reticulum, in neighbouring Dorado, lies this attractive 10th mag. spiral galaxy, NGC 1566. It is a type of galaxy known as a Seyfert galaxy with a bright, variable centre. *Anglo–Australian Telescope Board.*

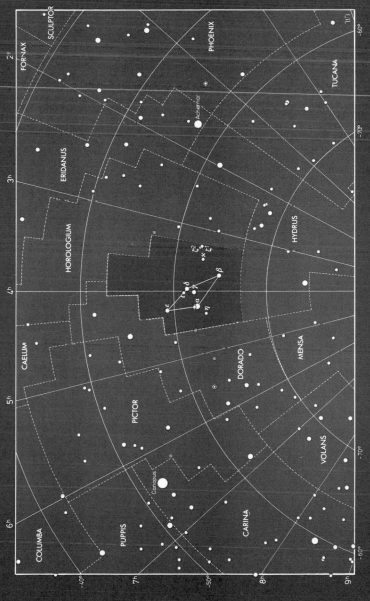

Magnitudes: −1 ● 0 ● 1 ● 2 ● 3 ● 4 ● 5 ● <5 ·
● ● Double or multiple   ◉ Variable
◉ Open cluster   ⊕ Globular cluster   ◯ Galaxy
▱ Diffuse neb.   ◇ Planetary neb.

SAGITTA   The Arrow

Despite its diminutive size – it is the 3rd-smallest constellation in the sky – this distinctive arrow-shaped group was known to the ancient Greeks. In the sky, the Arrow seems to be flying between Cygnus the Swan and Aquila the Eagle; one legend says the Arrow was shot by Hercules. Like its neighbour Vulpecula, Sagitta lies in a rich part of the Milky Way.

α (alpha) Sagittae, mag. 4.4, is a yellow giant 620 l.y. away.

β (beta) Sge, mag. 4.4, is an orange giant 650 l.y. away.

γ (gamma) Sge, mag. 3.5, the brightest star in the constellation, is an orange giant 170 l.y. away.

δ (delta) Sge, mag. 3.8, is a red giant 550 l.y. away.

ζ (zeta) Sge, 150 l.y. away, is a double star of mags. 5.0 and 8.8, divisible in small telescopes.

WZ Sge is a recurrent nova that flared up from 15th mag. to 7th mag. in 1913 and 1946; its location is worth checking in case of another flare-up.

M 71 (NGC 6838) is a dense 7th mag. cluster of stars 18,000 l.y. away, visible as a small misty patch in binoculars, and appearing nebulous in small telescopes. It is usually classed as a globular cluster, but some authorities regard it as a rich open cluster.

Scan northwards over the border from Sagitta into neighbouring Vulpecula with binoculars or telescope and you will encounter the Dumbbell Nebula, M 27, one of the showpiece sights of the sky. See page 254. *Hale Observatories photograph.*

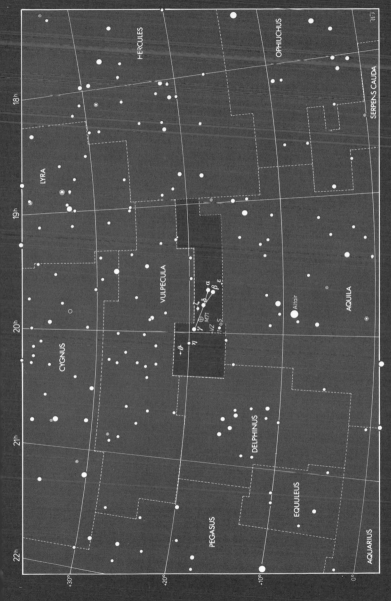

Magnitudes: −1 0 1 2 3 4 5 <5

●━● Double or multiple   ◉ ○ Variable

○ Open cluster   ⊕ Globular cluster   ○ Galaxy

▢ ◁ Diffuse neb.   ◇ Planetary neb.

SAGITTARIUS   The Archer

An ancient constellation depicting a centaur, half man, half beast, with a raised bow and arrow. Sagittarius has been visualized this way since at least ancient Greek times; it probably originated with the Sumerian civilization around the Euphrates, who saw him as Nergal, their archer god of war. It is an older constellation than the other celestial centaur, Centaurus, and is different in character. Whereas Centaurus is identified as a scholarly, beneficent creature, Sagittarius is depicted with a threatening look, aiming his arrow at the heart of Scorpius, the Scorpion. The bow is marked by the stars λ (lambda), δ (delta), and ε (epsilon) Sagittarii.

The Sun passes through the constellation from mid-December to mid-January, and thus lies in Sagittarius at the winter solstice, its farthest point south of the equator. The stars of Sagittarius are often visualized as forming a teapot, while a ladle shape known as the Milk Dipper is also drawn here – a suitable implement to dip into this rich region of the Milky Way. The centre of our Galaxy lies in Sagittarius, so that the Milky Way star fields are particularly rich here, as well as in the neighbouring Scutum and Scorpius. The actual centre of the Galaxy is marked by a radio and infrared source known as Sagittarius A, near the border with Scorpius. The main attraction of Sagittarius is its clusters and nebulae. Messier catalogued a total of 15 objects in Sagittarius, more than in any other constellation; only a selection of them can be mentioned here.

α (alpha) Sagittarii (Rukbat, knee, or Al Rami, the archer), mag. 4.0, is one of many examples in which the star labelled α (alpha) in a constellation is not the brightest. It is a blue-white star 200 l.y. away.

β (beta) Sgr consists of two unrelated stars seen separately by the naked eye. β^1 (beta^1) Sgr, called Arkab Prior (front part of the leg), is a blue-white star of mag. 3.9, 220 l.y. away; it has a mag. 7.3 companion visible in small telescopes. β^2 (beta^2) Sgr, Arkab Posterior (rear of the leg), is a white star, of mag. 4.3 and 130 l.y. away; it and β^1 (beta^1) Sgr appear in the same line of sight only by chance.

γ (gamma) Sgr (Al Nasl, point of the arrow), mag. 3.0, is a yellow giant 120 l.y. away.

δ (delta) Sgr (Kaus Meridionalis, middle of the bow), mag. 2.7, is an orange giant 82 l.y. away.

ε (epsilon) Sgr (Kaus Australis, southern part of the bow), mag. 1.9, is the brightest star in the constellation. It is blue-white and lies 85 l.y. away.

λ (lambda) Sgr (Kaus Borealis, northern part of the bow), mag. 2.8, is an orange giant 98 l.y. away.

σ (sigma) Sgr (Nunki), mag. 2.0, is a blue-white star 210 l.y. away. ▶

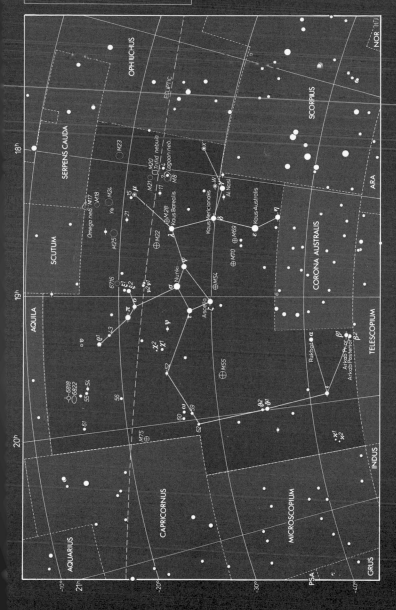

Magnitudes: −1 ● 0 ● 1 ● 2 ● 3 ● 4 ● 5 ● <5 ·
●–● Double or multiple   ◉ ○ Variable
○ Open cluster   ⊕ Globular cluster   ⬡ Galaxy
▱ Diffuse neb.   ✧ Planetary neb.

OPHIUCHUS

NOR

SERPENS CAUDA

SCORPIUS

ECLIPTIC

18ʰ

M23

Trifid nebula

Lagoon neb.

ARA

M21 M20

Omega neb. N17

M18

15 μ

M21 11

M8

· X

W

Kaus Meridionalis Al Nasl

Y φ

M24

21

M28

Kaus Borealis

Kaus Australis

δ

SCUTUM

M25

M22

λ

ε ⊕M69

γ

CORONA AUSTRALIS

67h6

φ

⊕M70

Nunki

AQUILA

ξ²ξ¹

ν²ν¹

Ascella ζ

⊕M54

π

σ

ρ¹

τ

ψ

6

19ʰ

ν

ρ²

χ¹

52

M55

TELESCOPIUM

β¹

α

β²

Rukbat

Arkab Prior

Arkab Posterior

6818

6822

55 54

56

60 ω 59

θ²

θ¹

61

M75 ⊕

62

χ¹

χ²

INDUS

20ʰ

CAPRICORNUS

MICROSCOPIUM

AQUARIUS

PSA

GRUS

21ʰ

−10°

−20°

−30°

−40°

Among the Milky Way starfields of Sagittarius lie two outstanding bright nebulae: M 20, the Trifid Nebula, top; and M 8, the Lagoon Nebula, below. To the top left of the Trifid Nebula is the loose cluster M 21. *Royal Observatory Edinburgh.*

◀ M 8 (NGC 6523), the Lagoon Nebula, is a famous gaseous nebula visible to the naked eye, encompassing the star cluster NGC 6530. M 8 is a fine object for binoculars or telescopes, covering an area twice the apparent size of the full Moon. A telescope shows a dark rift down the centre of the nebula. In the eastern half of the nebula is NGC 6530, a cluster of about 25 stars of mag. 7 and fainter, which have apparently formed recently from the surrounding gas. The other (western) side of the nebula is dominated by two main stars, the brighter of which is 6th mag. 9 Sagittarii. Long-exposure photographs show the nebula as an intense red, but visually it appears milky-white. M 8 is about 5000 l.y. away.

M 17 (NGC 6618), the Omega or Horseshoe Nebula, is another famous nebula, not as large or as bright as M 8. It is visible in binoculars on the border with Serpens. Small telescopes show it as an elongated smudge, while in larger instruments it appears arch-shaped, accounting for its popular name. M 17 lies 5000–6000 l.y. away. About 1° south of it is M 18 (NGC 6613), a loose cluster of about 12 stars of mag. 8 and fainter.

M 20 (NGC 6514), the Trifid Nebula, is a cloud of glowing gas far less impressive visually than on colour photographs. Even moderate-sized telescopes show it as only a diffuse patch of light centred on a double star of 7th and 8th mag. called HN 40, which has evidently been born from it. The Trifid Nebula gets its name from three dark lanes of dust that trisect it, as seen on long-exposure photographs. It is believed to be slightly more distant from us than M 8, which it lies 1.5° north of. In the same low-power field of view as M 20 lies the loose cluster M 21 (NGC 6531), consisting of about 50 stars between 8th and 12th mag.

M 22 (NGC 6656) is a rich globular cluster, one of the finest in the entire heavens and ranked 3rd only to ω (omega) Centauri and 47 Tucanae. Just visible to the naked eye, M 22 is easily seen in binoculars. It is a fine sight in small telescopes which reveal its noticeably elliptical outline. Telescopes of 75 mm aperture begin to resolve its outer regions, and larger apertures show its brightest stars to be reddish. Its nucleus is not as condensed as that of many other globulars. M 22 lies approx. 10,000 l.y. away.

M 23 (NGC 6494) is a rich, widespread cluster of 120 stars 4500 l.y. away, well seen in binoculars or small telescopes. It is elongated in shape, consisting of a field of 9th mag. stars of remarkably uniform appearance, some arranged in arcs.

M 24 is a rich and extensive Milky Way star field south of M 17 and M 18. Some observers restrict the name M 24 to a fainter and far smaller cluster of about 50 stars, known also as NGC 6603. The whole Milky Way star cloud in this region measures about 2° × 1° and is one of the most prominent parts of the Milky Way to the naked eye.

M 25 (IC 4725), 1800 l.y. away, is a scattered cluster of about 50 stars of mag. 6 and fainter. The cluster contains the Cepheid variable U Sgr, which varies from mags. 6.3–7.1 every 6 days 18 hrs.

M 55 (NGC 6809) is a nebulous-looking globular cluster visible in binoculars. Small telescopes resolve individual stars but show little central condensation of the cluster. M 55 lies 19,000 l.y. away.

SCORPIUS   The Scorpion

A resplendent constellation lying in a rich area of the Milky Way, and packed with exciting objects for users of small telescopes. In addition to the objects listed below, note three optical doubles – ζ (zeta), μ (mu), and ω (omega) Scorpii – formed by unrelated stars in the same line of sight. In mythology, Scorpius was the scorpion whose sting killed Orion. And in the sky Orion still flees from the scorpion, for Orion sets below the horizon as Scorpius rises. Scorpius clearly resembles the creature after which it is named, with a curve of stars forming its stinging tail. Its heart is marked by Antares, a name that means the rival of Mars, from its strong red colour. Northeast of β (beta) Scorpii, near the border with Ophiuchus, lies the brightest X-ray source in the sky, Scorpius X-1. This source has been identified with a 13th magnitude star, believed to be a close binary. Originally, in ancient Greek times and before, Scorpius was a much larger constellation, but the stars that once made up its claws have now been used to form the separate constellation of Libra. The Sun passes briefly through Scorpius during the last week of November.

α (alpha) Scorpii (Antares, rival of Mars), 330 l.y. away, is a red supergiant 300× the diameter of the Sun. It is a semi-regular variable, fluctuating between mags. 0.9–1.1 or fainter approximately every 5 years. Antares has a 6th mag. blue companion, so close that it requires at least 75 mm aperture and the steadiest of atmospheric conditions to be visible against the primary's glare.

β (beta) Sco (Graffias, crab), 540 l.y. away, is a striking double star divisible in the smallest telescopes, consisting of blue-white stars of mags. 2.6 and 4.9.

δ (delta) Sco (Dschubba, forehead), mag. 2.3, is a blue-white star 550 l.y. away.

ε (epsilon) Sco, mag. 2.3, is an orange giant 70 l.y. away.

θ (theta) Sco, mag. 1.9, is a yellow-white supergiant 910 l.y. away.

λ (lambda) Sco (Shaula, the sting), mag. 1.6, is a blue-white star 270 l.y. away.

ν (nu) Sco, 550 l.y. away, is a quadruple star similar to the famous Double Double in Lyra. A small telescope shows ν (nu) Scorpii as a wide double, with blue-white components of mags. 4.0 and 6.3. Telescopes of 75 mm and above reveal at high magnification that the fainter star is itself a close double of mags. 6.8 and 7.8. The brighter star is an even closer double of mags. 4.4 and 6.4, requiring an aperture of 150 mm to split.

ξ (xi) Sco, 85 l.y. away, is a celebrated multiple star. Small telescopes show it as a white star of mag. 4.2 with a mag. 7.2 orange companion; also visible in the same field is a fainter pair, called Struve 1999, composed of mag. 7.2 and mag. 8.0 stars, which are gravitationally connected to ξ (xi) Scorpii. Therefore, at first glance, ξ (xi) Sco looks like another Double Double. But the brightest star is itself a close double, consisting of two identical stars of mag. 4.9 orbiting each other every 46 years. At their widest in 1976 they were divisible in apertures of 100 mm, ▶

Magnitudes: -1 0 1 2 3 4 5 <5•
● ● Double or multiple   ◉ ○ Variable
○ Open cluster   ⊕ Globular cluster   ○ Galaxy
□ ⌐ Diffuse neb.   ◇ Planetary neb.

◄ but when closest around 1997 they will be impossible to split in amateur-sized telescopes.

M 4 (NGC 6121) is a large 7th mag. globular cluster visible in binoculars. In 100 mm telescopes, individual stars are resolved, and there is a noticeable bar of stars across the cluster's centre. M 4 is more loosely scattered than many globulars, and does not have a strong central concentration. M 4 is one of the closest of all globulars to us, lying 7500 l.y. away.

M 6 (NGC 6405) is an impressive 6th mag. star cluster easily seen in binoculars, covering approximately the same area of sky as the full Moon. It consists of a scattering of about 50 stars arranged in radiating chains. The brightest star of the cluster is a redgiant irregular variable called BM Sco, which varies between 6th–8th mag. M 6 lies about 1300 l.y. away.

M 7 (NGC 6475) is a large, brilliant star cluster visible to the naked eye, with an apparent diameter twice that of the Moon. Its 50 or so members are easily resolved in binoculars and small telescopes; the brightest are of mag. 6 and appear to be arranged in chains. M 7 lies 800 l.y. away.

M 80 (NGC 6093) is a small 8th mag. globular cluster visible in binoculars or a small telescope, appearing like the fuzzy head of a comet. It lies 36,000 l.y. away.

NGC 6231 is a large naked eye cluster of 120 or so stars, in a rich area of the Milky Way that is well worth sweeping with binoculars. The brightest stars of the group are of 6th mag., and give the impression of being a mini-Pleiades. The mag. 4.7 blue supergiant ζ^1 (zeta^1) Sco is an outlying member of this cluster, which lies 6000 l.y. away. NGC 6231 is connected to a larger, scattered cluster of fainter stars visible in binoculars, called H 12, which lies 1° to the north. The chain of stars linking NGC 6231 and H 12 outlines one of the spiral arms of our Galaxy.

Long-exposure photographs show that Antares and its surroundings are immersed in foggy nebulosity. In this photograph, Antares is lost in the bright cloud that it illuminates at bottom left. To its right is the globular star cluster M 4. Above right of M 4 a patch of nebulosity engulfs 3rd mag. σ (sigma) Scorpii. At the top of the picture are the foggy environs of ϱ (rho) Ophiuchi, from which complex smudges of bright and dark nebulosity extend southwards across the Ophiuchus-Scorpius border to Antares. *Royal Observatory Edinburgh.*

223

SCULPTOR The Sculptor

One of the faint and half-forgotten constellations introduced in the 1750s by Lacaille to fill in the southern celestial hemisphere. This constellation represents a sculptor's workshop, for reasons that remain obscure. Although its stars are of little interest, it has one point of note in that it contains the South Galactic Pole. Since Sculptor lies at right angles to the Milky Way, we can look out deep into space, unobscured by stars or gas, and see many faint galaxies. Among the galaxies in Sculptor is a faint dwarf elliptical member of our own Local Group, detectable only on long-exposure photographs taken with large telescopes.

α (alpha) Sculptoris, mag. 4.3, is a blue-white star 420 l.y. away.

β (beta) Scl, mag. 4.4, is a blue-white star 250 l.y. away.

γ (gamma) Scl, mag. 4.4, is a yellow giant 150 l.y. away.

δ (delta) Scl, mag. 4.5, is a white star 160 l.y. away.

ε (epsilon) Scl, 98 l.y. away, is a double star of mags. 5.3 and 9.4, visible in small telescopes.

κ¹ (kappa¹) Scl, 91 l.y. away, is a close double star with white components of mags. 6.2 and 6.3, at the limit of resolution in a 100 mm telescope.

R Scl is a deep red semi-regular variable star, which fluctuates between mags. 5.8–7.7 approx. every year.

NGC 55 is a 9th mag. spiral galaxy seen nearly edge-on, similar in size and shape to NGC 253 (see below), though not quite as bright. Its distance is estimated to be 6 million l.y.

NGC 253 is a 9th mag. spiral galaxy seen nearly edge-on, appearing cigar-shaped. Nearly 0.5° long, it can be picked up in binoculars, but requires at least 100 mm aperture to distinguish the central bulge and tightly wound arms. Its estimated distance is 9 million l.y. (See page 206.)

In addition to NGC 55 and NGC 253 in Sculptor, there is the much fainter (11th mag.) NGC 300, which photographs show to be a loose spiral. *Anglo-Australian Telescope Board.*

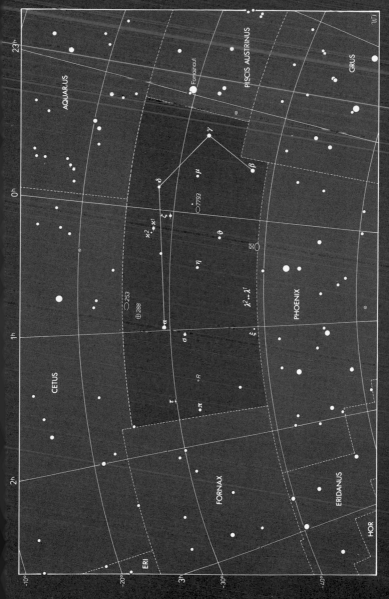

Magnitudes: -1 0 1 2 3 4 5 <5
●—● Double or multiple   ◉ ○ Variable
▢ Open cluster   ⊕ Globular cluster   ◯ Galaxy
▢ ⌒ Diffuse neb.   ◇ Planetary neb.

AQUARIUS

PISCES AUSTRINUS

GRUS

Fomalhaut

CETUS

PHOENIX

γ

μ

β

δ

7793

χ² χ¹

ζ

θ

5E

η

λ² λ¹

253

⊕ 288

α

σ

ξ

R

τ

π

FORNAX

ERIDANUS

HOR

ERI

23ʰ

0ʰ

1ʰ

2ʰ

3ʰ

-10°

-20°

-30°

-40°

SCUTUM The Shield

A faint constellation between Aquila and Serpens, introduced in 1690 by the Polish astronomer Johannes Hevelius under the title Scutum Sobieskii, Sobieski's Shield, in honour of his patron King John Sobieski III. Rich star clouds of the Milky Way are the main attraction in Scutum.

α (alpha) Scuti, mag. 3.9, is an orange giant 200 l.y. away.

δ (delta) Sct, 160 l.y. distant, is the prototype of a rare class of variable stars which pulsate in size every few hours, producing small amplitude brightness changes. δ (delta) Scuti itself varies from mags 4.7–4.8 every 4 hrs. 40 mins.

M 11 (NGC 6705), the Wild Duck Cluster, is a showpiece open cluster of about 200 stars, covering an area one third the apparent size of the Moon. It gets its name because of its noticeable fan shape, which resembles a flight of wild ducks. Binoculars show it as a misty patch, but in telescopes with magnifications of around × 100 it breaks up into a sparkling field of faint stardust. A slightly brighter, 8th mag. star lies at the fan's apex. M 11 is 5700 l.y. away.

Among the rich Milky Way starfields of Scutum lies M 11, the Wild Duck Cluster. In small instruments it appears noticeably fan-shaped. *Royal Observatory Edinburgh.*

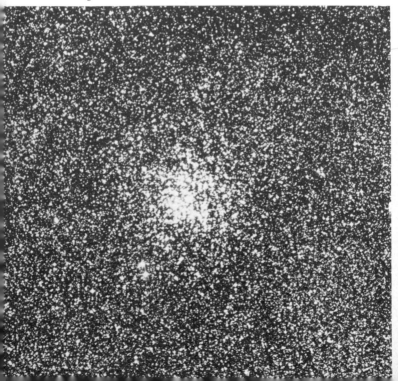

Magnitudes: -1 0 1 2 3 4 5 <5•

●—● Double or multiple ⊙ ○ Variable
○ Open cluster ⊕ Globular cluster ○ Galaxy
□ ⌐ Diffuse neb. ◇ Planetary neb.

HERCULES

SCORPIUS

OPHIUCHUS

ECLIPTIC

SERPENS CAUDA

17ʰ

18ʰ

19ʰ

SAGITTARIUS

ζ
α
γ
ε
β
R
M11 δ
M26

Altair

AQUILA

20ʰ

DELPHINUS

CAPRICORNUS

AQUARIUS

+10° 0° -10° -20°

SERPENS   The Serpent

An ancient constellation, representing a serpent wound around the body of the serpent bearer, who is represented by the adjacent constellation of Ophiuchus. In fact, Serpens is split into two halves, one either side of Ophiuchus: Serpens Caput, the head, which is the larger and more prominent half; and Serpens Cauda, the tail.

α (alpha) Serpentis (Unukalhai, serpent's neck), mag. 2.7, is an orange giant star 85 l.y. away.

β (beta) Ser, 120 l.y. distant, is a blue-white star in the serpent's head of mag. 3.7, with a 9th mag. companion visible in small telescopes. An unrelated 7th mag. star is visible nearby in binoculars.

γ (gamma) Ser, mag. 3.9, is a white star 39 l.y. away.

δ (delta) Ser, 170 l.y. away, is a white star of mag. 4.2 with a close mag. 5.2 companion visible in small telescopes at high magnification.

θ (theta) Ser (Alya), 100 l.y. distant, is an elegant pair of white stars of mags. 4.1 and 5.0, easily split in small telescopes.

ν (nu) Ser, 140 l.y. away, is a blue-white star of mag. 4.3 with a wide 9th mag. companion visible in binoculars or small telescopes.

τ (tau) Ser, mag. 5.2, is the brightest member of a loose cluster of eight stars of 6th mag. near β (beta) Serpentis, all visible in binoculars.

M 5 (NGC 5904) is a 6th mag. globular cluster 27,000 l.y. away, visible in binoculars or small telescopes. It is regarded as one of the finest globulars in the northern sky, second only to the famous M 13 in Hercules. Telescopes of 100 mm aperture or more reveal a brilliant, condensed centre and mottled outer regions, with apparent chains of stars radiating outwards. Close to M 5 lies 5 Serpentis, a 5th mag. star with a 10th mag. companion.

M 16 (NGC 6611) is an irregular, hazy-looking star cluster about 8000 l.y. distant, visible in binoculars or small telescopes. It consists of about 50 stars of 8th mag. and fainter, scattered over an area nearly equivalent to the apparent diameter of the Moon. The cluster is embedded in a nebula known as the Eagle. The Eagle Nebula is too faint to be well seen in amateur telescopes, but shows up beautifully on long-exposure photographs. (See page 230.)

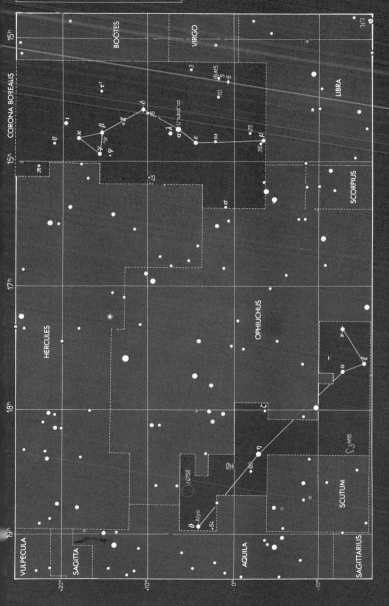

SEXTANS   The Sextant

A barren constellation south of Leo, introduced in the late 17th century by the Polish astronomer Hevelius. It commemorates the instrument he used for observing star positions. In fact, Hevelius continued to use his sextant for making naked eye sightings of star positions long after telescopes were available.

α (alpha) Sextantis, mag. 4.5, is a white star 330 l.y. away.

β (beta) Sex, mag. 5.1, is a blue-white star 520 l.y. away.

γ (gamma) Sex, mag. 5.1, is a white star 230 l.y. away.

δ (delta) Sex, mag. 5.2, is a white star 360 l.y. away.

NGC 3115 is a 9th mag. elliptical galaxy, 14 million l.y. away. Moderate-sized amateur telescopes show its elongated outline and brighter centre.

M 16, the Eagle Nebula in Serpens, is a visually spectacular combination of star cluster and gas cloud, with intruding fingers of cold, dark dust and gas. See description on page 228. *Hale Observatories photograph.*

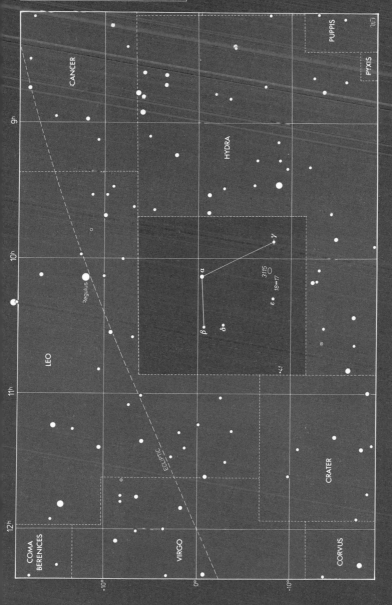

TAURUS   The Bull

One of the most ancient constellations. Taurus has been known to peoples throughout the world since the dawn of civilization, for the bull's attributes of strength and fertility mean that it has always held an honoured place in ceremony and religion. Usually only the head of the Bull is depicted, its face being formed by the V-shaped cluster of stars known as the Hyades. Its glinting red eye is marked by the star Aldebaran, and its long horns are tipped by the stars β (beta) and ζ (zeta) Tauri. In addition to the Hyades, Taurus contains the celebrated cluster of the Pleiades, the Seven Sisters. In Taurus occurred the famous supernova that was seen on Earth in 1054, which gave rise to the Crab Nebula, M 1. Between ε (epsilon) and ω (omega) Tauri lies the faint Hind's Variable Nebula, discovered last century by the English astronomer John Russell Hind; at the centre of this nebula lies the star T Tauri, prototype of a class of irregular variables believed to be stars in the process of formation. Each year the Taurid meteors radiate from a point near ε (epsilon) Tauri, reaching a maximum of about 12 meteors per hour on November 3. The Sun passes through the constellation from mid–May to late June.

α (alpha) Tauri (Aldebaran, the follower, i.e. of the Pleiades), mag. 0.9, is a red giant 68 l.y. away. Although it appears to be part of the Hyades cluster, it is in fact an unrelated foreground star.

β (beta) Tau (El Nath, the butting one), mag. 1.7, is a blue giant 140 l.y. away.

ζ (zeta) Tau, mag. 3.0, is a blue–white star 490 l.y. away.

θ (theta) Tau, 150 l.y. away, is a naked eye or binocular double in the Hyades, consisting of white and yellow giants of mags. 3.4 and 3.9.

κ (kappa) Tau, 75 l.y. away, is a white star of mag. 4.2 that forms a naked eye or binocular duo with an unrelated white star, 67 Tauri, of mag. 5.3, 120 l.y. away.

λ (lambda) Tau, 330 l.y. away, is an eclipsing binary of the Algol type, varying between mags. 3.5–4.0 every 4 days.

σ (sigma) Tau, 110 l.y. away, is a wide binocular double of white stars, mags. 4.7 and 5.1.

φ (phi) Tau, 290 l.y. away, is an optical double divisible in small telescopes, consisting of a mag. 5.0 orange giant and a white star of mag. 8.7.

χ (chi) Tau, 320 l.y. away, is a double star for small telescopes, with blue and gold components of mags. 5.4 and 8.2.

Hyades is a large and bright cluster of about 200 stars covering over 5° of sky. The brightest members form a noticeable V-shape, easily visible to the naked eye. In mythology, the Hyades were the daughters of Atlas and Aethra, and half-sisters of the Pleiades. Latest measurements place the Hyades at 150 l.y. from us; the distance of the Hyades is important, for it marks the first step in our distance

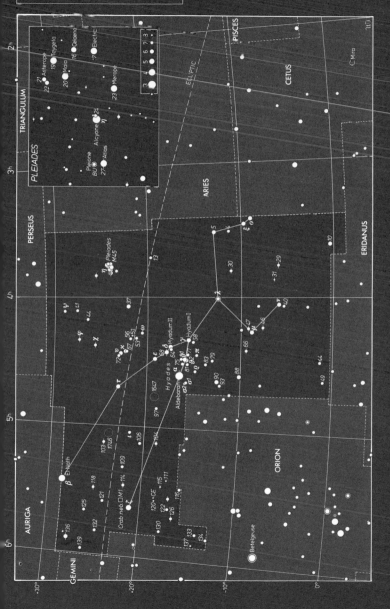

Magnitudes: −1 0 1 2 3 4 5 <5•

Double or multiple ● Variable

Open cluster ⊕ Globular cluster ◯ Galaxy

□ ⌒ Diffuse neb. ✧ Planetary neb.

PISCES

CETUS

Mira

TRIANGULUM

PLEIADES

21 22 Asterope 19 Taygeta 20 Maia 16 Celaeno 17 Electra 23 Merope

3 5 6 7 3

η 24 Alcyone BU Pleione 27 Atlas

PERSEUS

ARIES

ERIDANUS

ELLIPTIC

5 ξ ο

13

η Pleiades M45

10

30

31 29

ν ι 44 φ χ

37 56 55 53 51 ω

72 υ κ 67 62 ε γ δ Hyadum II Hyadum I 68

λ

40 ν

41 46 66

44

τ 1647 Hyades 64 66 61 θ1 75 81 π 77 58 90 92 83 79 88 49

Aldebaran α 54 θ2 σ1 93

97

ORION

1746 103 ι 106

β El Nath 118 121 109 114 120 CE 115 111 126 Crab neb □M1 7 116 ζ

125 132 136 139 130 133 134 137

Betelgeuse

AURIGA

GEMINI

−30° −20° −10° 0°

2h 3h 4h 5h 6h

TELESCOPIUM The Telescope

A constellation invented by the Frenchman Lacaille to honour the most important of astronomical instruments. As with so many of Lacaille's constellations it is faint and contrived, containing little to interest owners of small telescopes.

α (alpha) Telescopii, mag. 3.5, is a blue-white star 590 l.y. away.

δ (delta) Tel consists of two blue-white stars of mags. 5.0 and 5.1 visible separately in binoculars. They are unrelated to each other, being at 590 and 720 l.y. from us respectively.

ε (epsilon) Tel, mag. 4.5, is a yellow giant 190 l.y. away.

ζ (zeta) Tel, mag. 4.1, is a yellow giant 170 l.y. away.

◄ scale of the Galaxy. Because of its large size, the Hyades cluster is best studied with binoculars rather than with telescopes. The bright star Aldebaran is not a member of the cluster, but is superimposed on it by chance.

Pleiades, also known as M 45, is the brightest and most famous star cluster in the sky; it is popularly termed the Seven Sisters, after a group of mythological nymphs, the daughters of Atlas and Pleione. Approximately seven stars are visible to the naked eye, covering about 1° of sky; binoculars bring dozens more into view. A total of about 250 stars belong to the cluster, which lies 415 l.y. away. Unlike the stars of the Hyades, which are older and more evolved, the Pleiades formed within the last 50 million years, and include many young blue giants.

The brightest member of the Pleiades is η (eta) Tauri (Alcyone), mag. 2.9. Other prominent members are 16 Tau (Celaeno), mag. 5.5; 17 Tau (Electra), mag. 3.7; 19 Tau (Taygeta), mag. 4.3; 20 Tau (Maia), mag. 3.9; 21 Tau (Asterope), mag. 5.8; 23 Tau (Merope), mag. 4.2; 27 Tau (Atlas), mag. 3.6; and BU Tau (Pleione), a shell star that throws off rings of gas at irregular intervals, causing it to fluctuate unpredictably between mags. 5.0 and 5.5. The whole of the Pleiades is embedded in a faint nebulosity, the remains of the cloud from which the stars formed. This nebula is noticeable on long-exposure photographs, and under very clear conditions the brightest part of the nebula, around Merope, may be glimpsed in binoculars or small telescopes.

M 1 (NGC 1952) is the celebrated Crab Nebula, the remains of a star that exploded as a supernova. It was named in 1848 by Lord Rosse, who thought its shape as seen through his 183 cm (72 inch) telescope resembled a crab's pincer. Despite its fame it is a disappointing object for small telescopes, appearing as an elliptical 8th mag. wisp of nebulosity. At the centre of the nebula, beyond the reach of amateru telescopes, is a 16th mag. star, the remains of the star that exploded. This faint star is now known to be a pulsar.

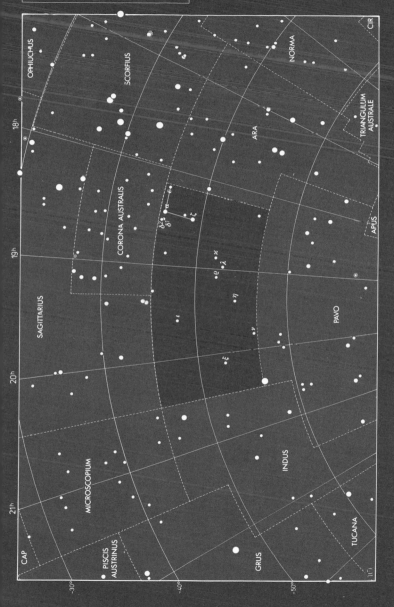

TRIANGULUM   The Triangle

A small but distinctive constellation lying between Andromeda and Aries, consisting of three main stars that form a thin delta shape; the Greeks referred to it as Deltoton. Its most important feature is the spiral galaxy M33, the 3rd-largest member of our Local Group of galaxies, following the Andromeda Galaxy and our own Milky Way.

α (alpha) Trianguli, mag. 3.4, is a white star 59 l.y. away.

β (beta) Tri, mag. 3.0, is a white star 110 l.y. away.

γ (gamma) Tri, mag. 4.0, is a blue-white star 150 l.y. away.

ι (iota) Tri, 280 l.y. away, is a mag. 4.9 golden yellow star with a close 7th mag. bluish companion visible in small telescopes.

M 33 (NGC 598) is a spiral galaxy 3.6 million l.y. away in our Local Group. Presented almost face-on, it covers an area of sky larger than the full Moon. Despite its size and proximity it is not prominent visually because of its low surface brightness. M 33 is best picked up in binoculars or small telescopes with low powers to enhance the contrast. Unlike most galaxies it does not have a noticeably stellar nucleus. Quite large amateur telescopes are needed to trace the spiral arms.

Spiral galaxy M 33 in Triangulum, a member of our Local Group. *Hale Observatories photograph.*

Magnitudes: −1 ● 0 ● 1 ● 2 ● 3 ● 4 ● 5 ● <5 ·
● − ● Double or multiple   ◉ ○ Variable
◇ Open cluster   ⊕ Globular cluster   ○ Galaxy
□ ⌒ Diffuse neb.   ✧ Planetary neb.

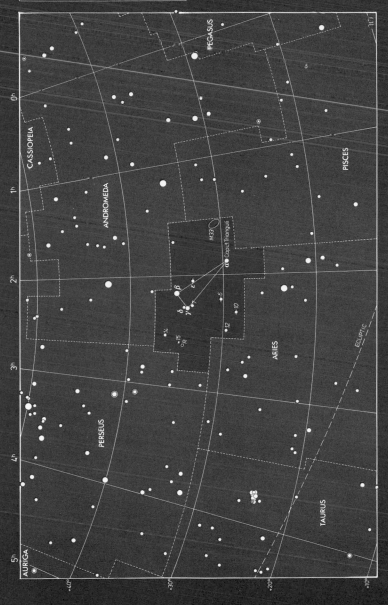

TRIANGULUM AUSTRALE The Southern Triangle

A small but readily distinguishable constellation near α (alpha) Centauri, introduced in 1603 by the German celestial cartographer Johann Bayer as a southern equivalent of the long-established northern Triangle, Triangulum.

α (alpha) Trianguli Australis, mag. 1.9, is an orange giant star 55 l.y. away.

β (beta) TrA, mag. 2.9, is a white star 33 l.y. away.

γ (gamma) TrA, mag. 2.9, is a blue-white star 91 l.y. away.

NGC 6025 is a binocular star cluster of about 30 members, of mag. 7 and fainter, 2000 l.y. away.

Johann Bayer (1572–1625)

Johann Bayer was a German lawyer of Augsburg and an amateur astronomer who, in 1603, published the first star atlas that covered the entire sky, the *Uranometria*. He based his coverage of the northern heavens on the observations made by the great Danish astronomer Tycho Brahe, while his information on the southern skies came from seafarers. In addition to the 48 constellations known since ancient times, Bayer introduced 12 new figures on his star maps to accommodate stars of the southern heavens: Apus, Chamaeleon, Dorado, Grus, Hydrus, Indus, Musca, Pavo, Phoenix, Triangulum Australe, Tucana and Volans. His most important legacy to astronomers was the introduction of the system of identifying stars by Greek letters. In each constellation the brightest stars were assigned letters of the Greek alphabet, usually (but not always) in approximate order of brightness. (Gemini, Orion and Sagittarius are famous examples of constellations in which the star labelled α (alpha) is not the brightest). Before Bayer's time, stars without proper names were identified by cumbersome descriptions such as 'in the left hand of the former twin', a reference to the 4th mag. star that we now call θ (theta) Geminorum. Clearly, to make sense of such descriptions astronomers had to know their constellation figures well, and even then there was considerable room for confusion. The system of so-called Bayer letters was a vast improvement, and remains in use to this day.

TRIANGULUM AUSTRALE
TrA · Trianguli Australis

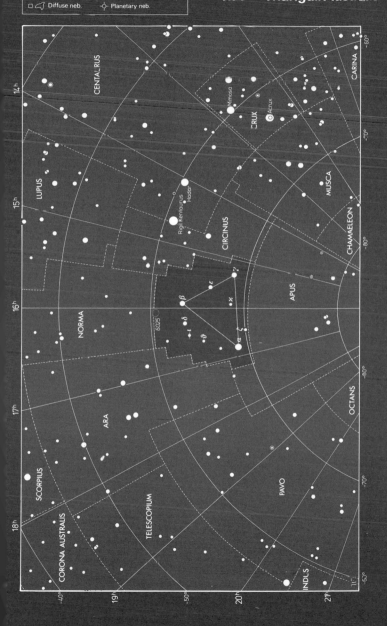

Magnitudes: -1 0 1 2 3 4 5 <5
- Double or multiple
- Variable
- Open cluster
- Globular cluster
- Galaxy
- Diffuse neb.
- Planetary neb.

CENTAURUS
CARINA
CRUX
Mimosa
Acrux
LUPUS
MUSCA
Rigil Kentaurus
Hadar
CHAMAELEON
CIRCINUS
APUS
NORMA
6025
β
ε
κ
δ
ι
θ
ζ
α
OCTANS
ARA
SCORPIUS
PAVO
CORONA AUSTRALIS
TELESCOPIUM
INDUS

TUCANA   The Toucan

A constellation near the south pole of the sky, introduced on the star chart of Johann Bayer in 1603. It represents a Toucan, the American bird with the large beak. Its most notable features are the Small Magellanic Cloud, and the globular cluster known as 47 Tucanae.

α (alpha) Tucanae, mag. 2.9, is an orange giant star 110 l.y. away.

β (beta) Tuc is a multiple star. Binoculars or small telescopes show that it consists of two almost identical blue-white stars, β^1 (beta[1]) and β^2 (beta[2]), of mag. 4.5, 150 l.y. away; nearby lies an unrelated mag. 5.1 third star, β^3 (beta[3]), 93 l.y. from us.

γ (gamma) Tuc, mag. 4.0, is a white star 150 l.y. away.

δ (delta) Tuc, mag. 4.5, is a blue-white star 250 l.y. away with a 9th mag. companion visible in small telescopes.

κ (kappa) Tuc, 59 l.y. away, is a double star consisting of components of mags. 5.1 and 7.3, visible in small telescopes. A wide 7th mag. star is also believed to be associated with the system.

47 Tuc (NGC 104) is a large and brilliant globular cluster, visible to the naked eye like a large fuzzy star – hence, on early charts, it was actually catalogued as a star. It is second only to ω (omega) Centauri as the largest and brightest globular cluster in the sky. Telescopes of 100 mm aperture begin to resolve 47 Tuc, and even binoculars show its brilliant central blaze. It is among the closest globulars to us, 19,000 l.y. away.

Small Magellanic Cloud (NGC 292) is a satellite galaxy of the Milky Way, as is its larger brother in Dorado. The Small Magellanic Cloud is visible to the naked eye as a nebulous, tadpole-shaped patch 3.5° across. Binoculars and small telescopes resolve it into clusters and glowing gas clouds. It lies 230,000 l.y. away. (For photograph, see page 158.)

NGC 362 is a globular cluster visible in binoculars near the edge of the Small Magellanic Cloud, but not associated with the Cloud. NGC 362 actually lies 40,000 l.y. away, in our own Galaxy.

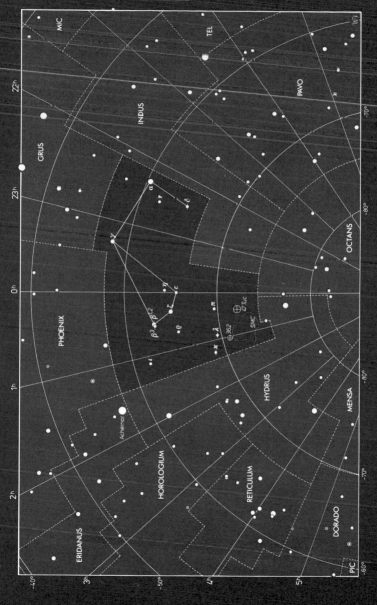

Magnitudes: -1 ● 0 ● 1 ● 2 ● 3 ● 4 ● 5 ● <5 ·
● Double or multiple   ◉ ○ Variable
Open cluster   ⊕ Globular cluster   ◇ Galaxy
□ Diffuse neb.   ◇ Planetary neb.

URSA MAJOR   The Great Bear

The 3rd-largest constellation in the sky. Its central feature is the seven stars that make up the shape variously called the Plough or the Big Dipper, undoubtedly the best known of all star patterns; references to it are found in writings dating back to the dawn of civilization (although why so many people, including the North American Indians, visualized this group as a bear remains a mystery). In Europe the pattern was seen as a wagon or chariot, and was associated by some people with the legendary King Arthur or with Charlemagne. Still others, notably the Arabs, viewed the dipper shape not as a bear but as a bier or coffin. The stars of the Big Dipper, except Benetnasch and Dubhe, are at similar distances from us, and form an associated group moving together through space. Two stars in the bowl of the Big Dipper, Merak and Dubhe, act as a pointer to Polaris, the Pole Star in the neighbouring constellation of Ursa Minor. The handle of the Big Dipper points in a curving arc to the bright star Arcturus in Boötes. Ursa Major contains the faint red dwarf Lalande 21185, magnitude 7.5, at 8.1 l.y. distance the Sun's 4th-closest stellar neighbour. Its name comes from its number in a catalogue by the 18th century French astronomer Joseph Lalande. Some astronomers suspect that Lalande 21185 has a planetary system. Ursa Major contains numerous faint galaxies, only a few of which are easily visible in amateur telescopes.

α (alpha) Ursae Majoris (Dubhe, the bear), mag. 1.8, is a yellow giant 75 l.y. away. It has a very close mag. 4.8 companion that orbits it every 44 years, and which requires a telescope of at least 220 mm aperture to be seen.

β (beta) UMa (Merak, loin), mag. 2.4, is a white star 62 l.y. distant.

γ (gamma) UMa (Phecda, thigh), mag. 2.4, is a white star 75 l.y. away.

δ (delta) UMa (Megrez, root of the tail), mag. 3.3, is a white star 65 l.y. away.

ε (epsilon) UMa (Alioth, meaning uncertain), 78 l.y. away, is a white star that varies between mags. 1.7–1.8 every 5.1 days.

ζ (zeta) UMa (Mizar, girdle), mag. 2.3, is a celebrated multiple star. Keen eyesight, or binoculars, reveal its mag. 4.0 companion Alcor. Mizar is 60 l.y. from Earth and Alcor is 80 l.y. away, too far apart to make this a genuine binary. A small telescope reveals that Mizar has another mag. 4.0 companion closer to it, which is related to it. This star was first seen by the Italian astronomer Giovanni Riccioli in 1650, making Mizar the first double star to be discovered telescopically. Mizar was also the first star discovered to be a spectroscopic binary, by the American astronomer E. C. Pickering in 1889. The companion of Mizar is another spectroscopic binary, as is Alcor, making this a complex multiple star system.

η (eta) UMa (Benetnasch or Alkaid, both from the Arabic for 'chief of the mourners'), mag. 1.9, is a blue-white star 160 l.y. away. ▶

M 81 is a beautifully perfect spiral galaxy in Ursa Major. *Hale Observatories photograph.*

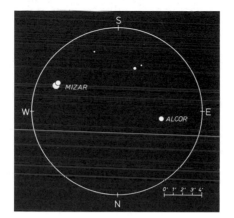

Mizar and Alcor as seen through a telescope. *Wil Tirion.*

◄ *ν* (nu) UMa, 150 l.y. away, is a mag. 3.5 orange giant with a mag. 9.7 greenish companion visible in small telescopes.

ξ (xi) UMa, 25 l.y. away, was the first double star to have its orbit computed. Its two yellow components (both spectroscopic binaries), of mags. 4.4 and 4.8, orbit each other every 60 years. Although at most times they are divisible in small telescopes, at their closest, around 1992, they require apertures of 150 mm to split them.

23 UMa, 82 l.y. away, is a white star of mag. 3.7 with a wide mag. 9.3 companion visible in small telescopes.

M 81 (NGC 3031) is a beautiful 8th mag. spiral galaxy, one of the brightest in the sky, visible in small telescopes as a roundish, softly glowing patch noticeably brighter towards the centre. It covers as much as half the apparent diameter of the full Moon. In the same telescopic field of view 0.5° to the north is M 82; the two galaxies are the same distance from us, 18 million l.y., and are related.

M 82 (NGC 3034) is a neighbour galaxy to M 81, although almost a magnitude fainter. It appears like an elongated blur, and is officially described as 'peculiar'. Long-exposure photographs show that it is actually an edge-on spiral galaxy interacting with a huge cloud of gas and dust.

M 97 (NGC 3587) is a large but dim planetary nebula known as the Owl Nebula, 2600 l.y. away.

M 101 (NGC 5457) is a famous spiral galaxy 23 million l.y. distant. Long-exposure photographs show it as a face-on galaxy with tightly wound spiral arms, but owners of small telescopes must be content with glimpsing it as a 9th mag. smudge, more than half the apparent diameter of the full Moon.

245

URSA MINOR   The Lesser Bear

A constellation said to have been introduced about 600 BC by the Greek astronomer Thales. Ursa Minor contains the present North Celestial Pole, within 1° of which lies the conveniently placed star we call Polaris (α (alpha) Ursae Minoris). Precession will bring Polaris closest to the pole around 2100 AD, after which it will start to move away again. Ursa Minor is also termed the Little Dipper because its seven brightest stars outline a shape like a smaller version of the Big Dipper in Ursa Major. The stars β (beta) and γ (gamma) Ursae Minoris in the bowl of the Little Dipper are termed the guardians of the pole. The stars in Ursa Minor are useful for checking atmospheric transparency, because their brightnesses run on one magnitude steps from 2nd–5th magnitude; the faintest member of Ursa Minor visible on a given night is a guide to the limiting magnitude on that night.

α (alpha) Ursae Minoris (Polaris or Cynosura) is a yellow supergiant star about 700 l.y. away. It is a Cepheid variable of small amplitude, varying in brightness between mags. 2.1–2.2 every 4 days. Polaris is also a double star, with a 9th mag. companion visible in small amateur telescopes.

β (beta) UMi (Kochab), mag. 2.1, is an orange giant 95 l.y. away.

γ (gamma) UMi (Pherkad), mag. 3.1, is a blue-white star 230 l.y. away. The 5th mag. star 11 UMi that appears near it, as seen by the naked eye or in binoculars, is unrelated.

δ (delta) UMi, mag. 4.4, is a blue-white star 140 l.y. away.

ε (epsilon) UMi, mag. 4.2, 200 l.y. away, is a yellow giant with a fainter companion that eclipses it every 39.5 days, causing a variation of about 0.1 of a mag., not discernable with the naked eye.

ζ (zeta) UMi, mag. 4.3, is a blue-white star 110 l.y. away.

η (eta) UMi, mag. 5.0, is a white star 91 l.y. away.

NGC 2841, a 10th mag. spiral galaxy in Ursa Major. *Hale Observatories photograph.*

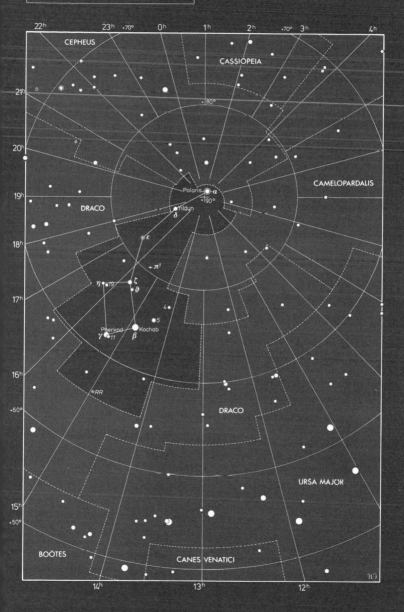

VELA   The Sails

Formerly a part of the ancient constellation of Argo Navis, representing the ship of Jason and the Argonauts, until that vessel was broken up by Nicolas Louis de Lacaille in the 1750s. Since Vela is only a subdivision of a once-larger constellation, it has no stars labelled α (alpha) or β (beta). The stars κ (kappa) and δ (delta) Velorum, in conjunction with ι (iota) and ε (epsilon) Carinae, form a shape known as the False Cross, because it is sometimes mistaken for the real Southern Cross. Vela lies in a part of the Milky Way rich with faint nebulosity, visible on long-exposure photographs. This nebulosity is known as the Gum Nebula, after the Australian astronomer Colin S. Gum who drew attention to it in 1952. The Gum Nebula is believed to be the remains of one or more super-novae that have occurred in Vela, the last of which, 6000 years ago, may have been seen by the ancient Sumerians at the dawn of civilization. Vela contains a pulsar that flashes 11 pulses per sec. In 1977 the Vela pulsar was seen flashing visually; it is the faintest star ever identified optically.

γ (gamma) Velorum, 650 l.y. away, is an interesting multiple star. Binoculars or small telescopes show that it consists of two blue-white components of mags. 1.8 and 4.3. The brighter of these is the brightest known Wolf-Rayet star, a rare class of stars with very hot surfaces which seem to be ejecting gas. There are also two wider companions, of 9th and 10th mag.

δ (delta) Vel is a white star 68 l.y. away of mag. 2.0 with a close mag. 6.5 companion, requiring telescopes of 150 mm aperture or above.

λ (lambda) Vel, mag. 2.2, is a yellow supergiant 490 l.y. away.

μ (mu) Vel, mag. 2.7, is a yellow giant 98 l.y. away.

κ (kappa) Vel, mag. 2.5, is a blue-white star 390 l.y. away.

NGC 2547 is a cluster of about 50 stars, just visible to the naked eye and best seen in binoculars, lying 3100 l.y. away.

IC 2391 is a bright cluster of 20 stars, 850 l.y. away, scattered around the 4th mag. star o (omicron) Velorum.

IC 2395 is a binocular cluster of 16 stars of mag. 9 and fainter, 4300 l.y. away. Also visible to the south is the 9th mag. cluster NGC 2670.

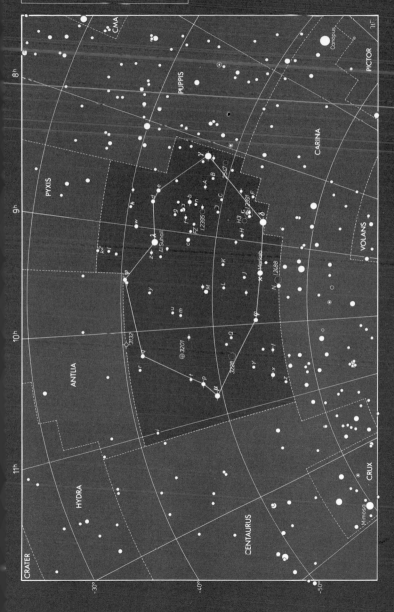

VIRGO   The Virgin

The 2nd-largest constellation in the sky. Many myths are associated with this constellation, which has usually been seen as a beautiful and virtuous maiden. We know her as Astraea, the Roman goddess of justice, with neighbouring Libra representing her scales of justice. But other legends have associated Virgo with successful harvests, and so she is pictured holding an ear of wheat (the star Spica) in her left hand and a palm leaf in her right hand. In these legends she is identified with the Greek harvest goddess Demeter (the Roman Ceres) or more usually with her daughter Persephone (the Roman Proserpine). The Sun passes through the constellation from mid-September to early November, and thus is within Virgo's boundaries at the time of the autumnal equinox around September 23 each year, when the Sun moves south of the celestial equator. Virgo contains a rich cluster of galaxies (the nearest major galaxy cluster to us), which spills over into neighbouring Coma Berenices; the area is often known as the realm of the galaxies. The Virgo Galaxy Cluster lies about 65 million l.y. away, and contains about 3000 members, several dozen of which are visible in apertures of 150 mm or so, although they appear as little more than hazy patches of light. Some of the brightest members of the cluster are mentioned below. Virgo contains the brightest quasar, 3C 273, located about 5° north of γ (gamma) Virginis. It is unrelated to the Virgo Galaxy Cluster. Optically, 3C 273 appears as a 13th magnitude blue star. It is estimated to lie about 3000 million l.y. away.

α (alpha) Virginis (Spica, ear of wheat), mag. 1.0, is a blue-white star 260 l.y. away. It is an eclipsing binary, varying by about 0.1 of a mag. every 4 days.

β (beta) Vir, mag. 3.6, is a yellow star 33 l.y. away.

γ (gamma) Vir (Porrima, named after a Roman goddess of prophecy), 36 l.y. away, is a celebrated double star. Together, the stars shine as mag. 2.8. But small telescopes reveal γ (gamma) Virginis to consist of a matching pair of yellow-white stars each of mag. 3.6. They orbit every 172 years, and at present appear to be closing together. By the year 2000 they will need a 75 mm telescope to split them, and at their closest around 2008 they will be indivisible in amateur telescopes.

δ (delta) Vir, mag. 3.4, is a red giant 180 l.y. away.

ε (epsilon) Vir (Vindemiatrix, grape gatherer), mag. 2.8, is a yellow giant 100 l.y. away.

θ (theta) Vir, 140 l.y. away, is a double star visible in small telescopes, consisting of blue-white components of mags. 4.4 and 8.6.

τ (tau) Vir, mag. 4.3 and 100 l.y. away, forms a wide optical double for small telescopes with a mag. 9.5 companion.

φ (phi) Vir, 95 l.y. away, is a yellow giant of mag. 4.8 with a mag. 9.2 orange ▶

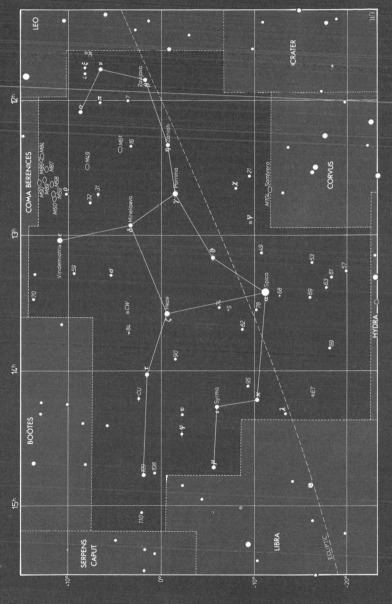

VOLANS   The Flying Fish

This figure, on the borders of Carina, was introduced on the 1603 star map of Johann Bayer under the title Piscis Volans. None of its stars are particularly bright, but there are two fine double stars for small telescopes.

α (alpha) Volantis, mag. 4.0, is a white star 78 l.y. away.

β (beta) Vol, mag. 3.8, is an orange giant 190 l.y. away.

γ (gamma) Vol, 130 l.y. away, is a pair of white and yellow stars of mags. 3.8 and 5.7, divided in small telescopes.

δ (delta) Vol, mag. 4.0, is a yellow supergiant 2400 l.y. away.

ε (epsilon) Vol, 390 l.y. away, is a mag. 4.4 blue star with a mag. 8 companion visible in small telescopes.

◄ companion, difficult to see in the smallest telescopes because of the magnitude contrast.

M 49 (NGC 4472) is a 9th mag. elliptical giant galaxy visible in 75 mm telescopes under low power. It is one of the brightest members of the Virgo Galaxy Cluster.

M 58 (NGC 4579) is a 9th mag. spiral galaxy with a noticeably brighter core.

M 60 (NGC 4649) is a 9th mag. giant elliptical galaxy, one of the most prominent members of the Virgo Cluster and detectable with 75 mm aperture. About a quarter the way from M 60 to M 58 lies M 59 (NGC 4621), a 10th mag. elliptical galaxy.

M 84 (NGC 4374) and M 86 (NGC 4406) are a pair of 9th mag. elliptical galaxies both appearing in the same telescopic field as fuzzy patches with noticeably brighter cores.

M 87 (NGC 4486) is a celebrated giant elliptical galaxy. It is a radio source, known as Virgo A, and an X-ray source. Photographs with large telescopes show a jet of matter emerging from M 87, as though ejected in an explosion. In amateur telescopes, M 87 appears as a rounded, 9th mag. glow with a noticeable nucleus.

M 104 (NGC 4594) is a 9th mag. spiral galaxy seen edge-on. It is popularly known as the Sombrero Hat because of its distinctive appearance on long-exposure photographs which show it with a bulging nucleus ringed by spiral arms. A dark lane of dust around its rim, revealed on photographs, needs apertures above 150 mm to be detected visually. It is about 35 million l.y. away.

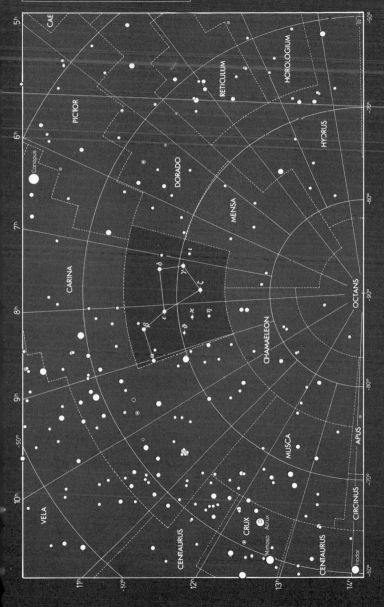

VULPECULA The Fox

A faint constellation at the head of Cygnus. It originated in 1690 with the Polish astronomer Johannes Hevelius, who called it Vulpecula cum Anser, the Fox and Goose. Since then, the goose has fled, leaving the fox. In 1967 this diminutive constellation was the site of an astounding discovery – the first pulsar, or flashing radio source, which was detected by radio astronomers in Cambridge, England. On the border of Vulpecula with Sagitta is a notable little group of 6th and 7th magnitude stars known colloquially as the Coathanger, a striking scene in binoculars. The Coathanger's most remarkable feature is an almost straight line of six stars; a curve of stars, forming the Coathanger's hook, extends from the centre of this line.

α (alpha) Vulpeculae, mag. 4.4, is a red giant 270 l.y. away. An unrelated 6th mag. star, 8 Vul, is seen to accompany it in binoculars.

M 27 (NGC 6853), the Dumbbell Nebula, is a large and bright planetary nebula, reputed to be the most conspicuous of its kind, visible in binoculars but seen better in telescopes. Of 8th mag., M 27 covers $\frac{1}{4}$ of the diameter of the full Moon. Visually it appears a dumbell-shaped, misty green glow. Long-exposure photographs show a complex pattern of blue and pink gas. M 27 is 1250 l.y. away.

NGC 2442, a 12th mag. barred spiral galaxy in Volans. *Anglo-Australian Telescope Board.*

The Pleiades (the Seven Sisters, the daughters of Atlas) is a famous open star cluster in the constellation Taurus, embedded in faint nebulosity. (See page 232.) *Colin Hunt.*

SECTION II   Stars

Stars are balls of gas made incandescent by energy from nuclear reactions deep in their interiors. They come in a wide range of sizes and brightnesses, from faint dwarfs one per cent the Sun's diameter to dazzling supergiants hundreds of times the size of the Sun. They range in temperature from intensely hot blue-white stars (more than 20,000°C) to cool red stars (3000°C). The Sun, which is a medium-temperature yellow star, turns out to be pretty average in all respects.

Stars are born from massive clouds of gas and dust within our Galaxy. An interstellar gas cloud is termed a *nebula* (plural: nebulae), from the Latin meaning 'cloud'. A nebula is not uniformly distributed in space, but contains denser knots – the seeds of future stars. If the knot is dense enough it begins to contract under the inward pull of its own gravity. As it gets smaller and denser it heats up, until conditions of temperature and pressure at the centre of the shrinking blob become so extreme that nuclear reactions begin. The gas blob has switched on to become a true star, generating its own heat and light for millions of years.

Several star-spawning clouds are well within reach of observation by amateurs. Most famous is the Orion Nebula, marking the sword in the constellation of Orion the Hunter. This nebula is visible as a hazy green glow to the naked eye; binoculars show it more clearly. At the centre of the Orion Nebula is a star called θ^1 (theta^1) Orionis, which small telescopes show consists of four component stars. Energy emitted by the brightest of these four stars makes the nebula shine. But behind the bright, visible part of the cloud is an even larger, still-dark area where stars are being born at this moment. The Orion Nebula is estimated to contain enough matter to produce 10,000 stars: it is a star cluster in the making. Another famous stellar birthplace is the Tarantula Nebula in the southern constellation Dorado, which dwarfs the Orion Nebula and is in fact the largest nebula known.

One celebrated young cluster of stars is the Pleiades in the constellation Taurus, the Bull. Long-exposure photographs show the stars of the Pleiades to be still surrounded by wisps of the nebula from which they formed. At least five members of the Pleiades can be distinguished by normal eyesight; binoculars and small telescopes bring dozens more members into view. The whole cluster is estimated to contain about 250 stars. The brightest and youngest of these formed no more than two million years ago, making them extremely youthful by astronomical standards.

Left : the Jewel Box, NGC 4755 (also known as κ (kappa) Crucis), a glittering star cluster in the southern constellation Crux, the Southern Cross. Most of its stars are blue-white, but at its centre is a red giant. *Anglo–Australian Telescope Board.*

Below : The spidery shape of the Tarantula Nebula, NGC 2070, glows brightly in the constellation of Dorado. The nebula is part of the Large Magellanic Cloud, as are the other nebulae visible near to it in this photograph. *Royal Observatory Edinburgh.*

A true-colour photograph of the Orion Nebula, M 42, whose extensive swirls of ghostly glowing gas make it one of the most celebrated objects in the entire heavens. *Anglo-Australian Telescope Board.*

The Pleiades is an example of a type of cluster referred to as an *open cluster* or *galactic cluster*. About 1000 are known to astronomers, and many of them are listed in this book. Near to the Pleiades in Taurus is a larger and older open cluster, the Hyades, which is estimated to be about 500 million years old. Being older than the Pleiades, the stars have had more chance to drift apart. Eventually, most open clusters disperse completely. The Sun was probably a member of such a cluster when it was born 4600 million years ago. A different type of cluster is a globular cluster, dealt with on page 278.

Nebulae are made of a 10:1 mixture of hydrogen and helium, the primary constituents of the Universe, so stars naturally enough have the same composition. Stars get their energy from nuclear reactions which transform hydrogen into helium. In the reaction, four hydrogen atoms are crushed together to make one atom of helium; an uncontrolled version of the same reaction occurs in a hydrogen bomb.

There are certain limits on the size of a star. A gas blob with less than six per cent the Sun's mass cannot become a star, because conditions in its interior will not become sufficiently extreme for nuclear reactions to begin. This six per cent limit can be considered the dividing line between a planet and a star. If the gaseous planet Jupiter in our solar system had been about 60 times more massive than it actually is, it would have become a small star. At the other end of the scale, the largest stars have masses of about $100 \times$ that of the Sun. Until recently it was thought that stars more massive than this would produce so much energy that they would literally disintegrate, but this may not be true in all cases. A few stars are known that seem to have masses greater than 100 Suns, one example being the peculiar star η (eta) Carinae.

A star's most vital statistic is its mass, for this factor affects everything else about it: its temperature, its brightness and its lifetime. The stars with the least mass are, not surprisingly, the coolest; they are known as *red dwarfs*. A typical red dwarf such as Barnard's Star, the second closest star to the Sun, has a mass about a tenth that of the Sun and glows a dull red with a surface temperature of about 3000°C. Even though Barnard's Star is only six light years away, it is too faint to be seen with the naked eye. Surprisingly enough, stars with the lowest mass live the longest. Their nuclear fires burn so slowly that they can survive for as much as a million million years, 100 times longer than the Sun. The Sun itself, which by definition is of one solar mass, has a surface temperature of 5500°C and is expected to live for about 10,000 million years. It is currently in the prime of its life.

Moving up the scale, a star such as Sirius, which is twice the Sun's mass, can live for only about 1000 million years, a tenth of the Sun's age. The surface temperature of Sirius is a blue-white 11,000°C. Larger and hotter still, the star Spica in the constellation Virgo has a mass of about 11 Suns and a surface temperature of around 24,000°C. The

etime of this intensely hot, highly luminous star is less than one per
ent of the lifetime of the Sun.

A star's colour is a direct indicator of its temperature. The most precise way to measure a star's temperature is by studying the spectrum of
its light, which is done by splitting the light up in a device called a
spectroscope. Stars are classified into a sequence of so-called *spectral
types* according to their temperature. The bluest and hottest stars are
classified as spectral types O and B (Spica is a B-type star). Then come
the cooler blue-white A-type stars, which include Sirius, and then F-
type stars which appear white; Procyon is an F-type star. G-type stars
appear yellow-white; they include the Sun, α (alpha) Centauri and
τ (tau) Ceti. Cooler still are K-type stars, such as ε (epsilon) Eridani,
which appear orange. Coolest of all are the red stars with an M-type
spectrum, of which Barnard's Star is an example. Each spectral type is
subdivided into 10 steps from 0–9; on this more precise scale, the Sun
ranks as a G2 star.

Stellar Spectral Types

Type	Colour	Temperature range ($°C$)	Examples
O	Blue	40,000–25,000	Zeta Puppis (supergiant)
B	Blue	25,000–11,000	Spica (main sequence) Regulus (main sequence) Rigel (supergiant)
A	Blue-white	11,000–7500	Vega (main sequence) Sirius (main sequence) Deneb (supergiant)
F	White	7500–6000	Canopus (supergiant) Procyon (subgiant) Polaris (supergiant)
G	Yellow-white	6000–5000	Sun (main sequence) Alpha Centauri (main sequence) Tau Ceti (main sequence) Capella (giant)
K	Orange	5000–3500	Epsilon Eridani (main sequence) Arcturus (giant) Aldebaran (giant)
M	Red	3500–3000	Barnard's Star (main sequence) Antares (supergiant) Betelgeuse (supergiant)

The largest and most beautiful planetary nebula, the Helix Nebula, NGC 7293 in Aquarius, is a shell of gas thrown off by a dying star. The red colour of its outer part is caused by nitrogen and hydrogen, the blue-green of the central part by oxygen. *Anglo-Australian Telescope Board.*

The seemingly haphazard lettering sequence for the spectral types is the result of a previous classification scheme which was rearranged and shortened to produce the present system. The sequence of stellar spectral types is remembered by the mnemonic: 'Oh Be A Fine Girl, Kiss Me'.

When the spectral type of stars is plotted on a graph against their actual luminosity (absolute magnitude), all stars that are in stable,

Right: The Trifid
Nebula, M 20 in
Sagittarius, is a cloud of
gas and dust in which
stars are forming. In one
part of the nebula the light
from new-born stars
makes hydrogen gas glow
pinkish red. Another part
of the nebula appears
blue, caused by starlight
reflected from dust grains.
Dark lanes of dust trisect
the nebula, giving rise to
its popular name.
*Anglo-Australian
Telescope Board.*

Right: M 16 in Serpens is
a cluster of stars embedded
in a faint nebula called the
Eagle. The cluster is easily
seen in amateur instru-
ments, but the nebula
shows up well only on
long-exposure photo-
graphs such as this.
*Anglo-Australian
Telescope Board.*

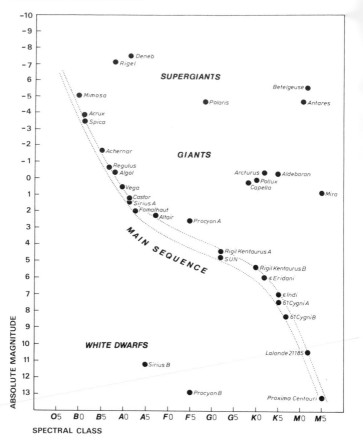

In the Hertzsprung-Russell diagram, the actual brightness of stars (their absolute magnitude) is plotted against their temperature (spectral class). Stars of different types fall in different areas of the Hertzsprung-Russell diagram. The band running from top left to lower right is the main sequence, on which the Sun falls. Giants and supergiants lie above the main sequence; white dwarfs lie below and to the left of it. *Wil Tirion.*

hydrogen-burning middle age lie in a well-defined band across the graph known as the main sequence. A star's position along the main sequence is fixed by its mass, with the least massive stars being at the bottom end and the most massive stars at the top end. The Sun, as befits its middle-of-the-road nature, lies about halfway along the main sequence. Such a plot of star brightness against spectral type is known as a Hertzsprung-Russell diagram, after the Danish astronomer Ejnar Hertzsprung and the American Henry Norris Russell who devised it in 1911–13.

Although most stars lie on the main sequence of the Hertzsprung-Russell diagram, a number of particularly bright stars lie above and to the right of the main sequence, while a few faint stars lie below and to the left of the main sequence. These stars are all in late stages of evolution. We can best understand what is happening to them by following the future evolution predicted for the Sun.

As we have seen, the Sun formed about 4600 million years ago and is about halfway through its expected lifespan. In a few thousand million years, though, it will start to run out of hydrogen at its core. In search of more hydrogen to use as fuel, the nuclear reactions inside the Sun will start to move outwards, releasing more energy. Eventually, when surrounded by a shell of burning hydrogen, even the helium in the Sun's core will enter into nuclear reactions of its own, fusing together to form carbon. With all this extra energy being given off, the Sun will become much brighter than it is today and will start to swell alarmingly in size. But as the Sun's outer layers expand they will also cool, becoming redder in colour, so that the Sun turns into a red giant similar to the bright stars Aldebaran and Arcturus. At its largest, the red giant Sun will grow to at least 100 times its present diameter, engulfing the Earth within its distended outer layers. Needless to say, all life on our planet will long since have become extinct.

On the Hertzsprung-Russell diagram, the Sun's increase in brightness will move it upwards off the main sequence, and its change in spectral type will in addition move it to the right. Stars at the top end of the main sequence become so big and bright at this stage of their evolution that they are referred to not as mere giants but as supergiants. Prominent examples of red supergiants are Betelgeuse and Antares, both of which are hundreds of times larger than the Sun. Other stars which have not yet evolved far enough to become red in colour but are

Stellar Luminosity Classes

Ia	Bright supergiant
Iab	Less bright supergiant
Ib	Supergiant
II	Bright giant
III	Giant
IV	Subgiant
V	Main sequence
VI	Subdwarf
VII	White dwarf

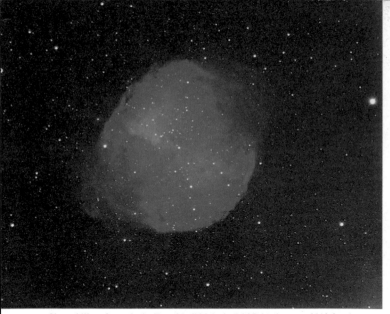

Shaped like a figure 8, the Dumbbell Nebula, M 27, is a large and bright planetary nebula in Vulpecula. Photograph by Laird Thompson with the 3.6 m Canada-France-Hawaii Telescope. *University of Hawaii.*

M 57, the Ring Nebula in Lyra, photographed by Laird Thompson with the Canada-France-Hawaii Telescope. The red colour is produced by hydrogen; the yellow and blue comes from oxygen. *University of Hawaii.*

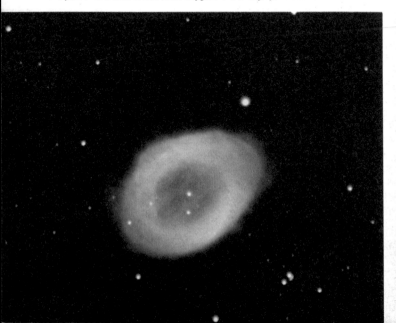

The Horsehead Nebula, a cloud of dark dust looking in silhouette like an interstellar chess piece against the redness of glowing hydrogen gas. At left is the overexposed image of ζ (zeta) Orionis. *Royal Observatory Edinburgh.*

Crab Nebula, M 1: remains of a star that exploded as a supernova. The blue colour is given by electrons spiralling in the Crab's intense magnetic field. Canada-France-Hawaii Telescope photograph by Laird Thompson. *University of Hawaii.*

nevertheless firmly in the supergiant bracket are Rigel, Deneb and Canopus.

To distinguish whether a star is, say, a giant, a supergiant, or lies on the main sequence, astronomers assign stars a luminosity class (see table) in addition to the spectral type. (Note that astronomers tend to think of stars as being either dwarfs or giants, depending whether they are still on the main sequence or have evolved off it. Main sequence stars are frequently referred to as dwarfs, even though the most massive of them may be several times larger than the Sun.) Taken together, the spectral type and luminosity class define the main properties of a star as it exists at present. But those properties change as the star ages.

Stars spend only a few per cent of their total lifetime in the red giant phase, which in the case of stars like the Sun amounts to no more than a few hundred million years. A red giant is a star that has grown old and is about to die. Once a red giant has swollen to its maximum size, its distended outer layers drift off into space, forming a stellar smoke ring known somewhat confusingly as a *planetary nebula*, even though it has nothing to do with planets. The name was given in 1785 by William Herschel who said that they looked like the small, rounded disks of planets as seen through his telescope. Probably the most celebrated of all planetary nebulae is the Ring Nebula in Lyra, though it is not the easiest to see. Much larger is the Dumbbell Nebula in Vulpecula, which can be picked up in binoculars on a clear, dark night. Two small but bright planetary nebulae for amateur telescopes are NGC 6826 in Cygnus and NGC 7662 in Andromeda.

At the centre of a planetary nebula, the core of the former red giant is exposed as a small, intensely hot star. Once the surrounding gases of the planetary nebula have dispersed, usually after thousands of years, the central star remains as a so-called *white dwarf*. A white dwarf is only about the diameter of the Earth, but contains most of the matter of the original star; only about 10 per cent of the star's mass is lost in the planetary nebula stage. White dwarfs are therefore exceptionally dense bodies. A teaspoonful of white dwarf material would have a mass of thousands of kilograms. Over thousands of millions of years, white dwarfs slowly cool off and fade into oblivion.

Being so small, white dwarfs are very faint. Not one is visible to the naked eye. Both the nearby bright stars Sirius and Procyon have white dwarf companions, but Procyon's companion is too close to its parent to be distinguishable in amateur telescopes, and the companion of Sirius can only be glimpsed under the most favourable conditions. The easiest white dwarf to see is a companion of the star o^2 (omicron^2) Eridani (also known as 40 Eridani); small telescopes show it. Of added interest is a fainter third member of this system, a red dwarf, also visible in amateur telescopes.

Our Sun, it seems, is destined to go through the stage of being a planetary nebula before fading out as a white dwarf. But stars with several times the Sun's mass, towards the top end of the main sequence, suffer a far more spectacular end. As we have seen, they first become dazzling supergiants rather than mere giants. They do not get a chance to reach the planetary nebula stage. So massive are they that the nuclear reactions at their centres continue in runaway fashion until the star becomes unstable and explodes. Such an explosion is known as a *supernova*.

In a supernova eruption a star's brightness increases millions of times, so that for a few days the star can rival the brilliance of an entire galaxy. The shattered outer layers of the star are thrown off into space at speeds of around 5000 km per second. In 1054 astronomers on Earth saw a star erupt as a supernova in the constellation of Taurus. The star became brighter than Venus and was visible in daylight for three weeks. It finally faded below naked-eye visibility more than a year after it had first appeared.

At the site of that explosion lies one of the most famous objects in the heavens: the Crab Nebula, the shattered remains of the star that erupted as a supernova. The Crab Nebula is visible as a smudgy patch in amateur telescopes, but is best seen on long-exposure photographs taken with large instruments. Over the next 50,000 years or so the gases of the Crab Nebula will disperse into space, forming delicate traceries like those of the Veil Nebula in Cygnus, itself the remains of a former supernova.

The last supernova observed in our Galaxy was in 1604. This star, in the constellation Ophiuchus, reached a maximum magnitude of just over −2, as bright as Jupiter. It was studied by the German astronomer Johannes Kepler and is often known as Kepler's Star in his honour. Hundreds of supernovae have been seen by telescope in other galaxies since then; another supernova in our Galaxy is long overdue. When it comes, it should be a spectacular sight. Many astronomers dream of seeing the next supernova outshining the other stars in the night sky, dazzling the naked eye and bright enough to cast shadows.

A star may not blow itself entirely to bits in a supernova explosion. Sometimes the central core of the exploded star is left as an object even smaller and denser than a white dwarf, known as a *neutron star*. In a neutron star, the protons and electrons of the star's atoms have been crushed by the tremendous forces of the supernova so that they combine to form the particles known as neutrons. A typical neutron star is a mere 20 km in diameter, but contains as much mass as one or two Suns. Being so minute, neutron stars can spin very rapidly without flying apart. Each time they spin we see a flash of radiation like a lighthouse beam. Astronomers have detected radio pulses from several hundred such sources which they term *pulsars*; one lies at the centre of the Crab Nebula. The Crab pulsar radiates 30 times per second; others pulse

The η (eta) Carinae Nebula, NGC 3372: a large, bright cloud of glowing hydrogen gas containing clusters of hot young stars. The nebula is named after η (eta) Carinae itself, a massive star that lies in the brightest central part of the nebula, near to a dark notch called the Keyhole. η (eta) Carinae could become a supernova within the next 10,000 years. *Royal Observatory Edinburgh*.

Twisted filaments of gas veining the constellation Vela are the dismembered remains of a star that erupted in a supernova explosion about 10,000 years ago. The core of the star was left behind as a faint pulsar; this flashes 11 times per second. The Vela supernova remnant is part of the much larger Gum Nebula, named after the astronomer Colin S. Gum. *Royal Observatory Edinburgh.*

more slowly, down to once every four seconds. Most neutron stars are too faint to be seen optically, but the pulsar in the Crab Nebula has been seen flashing in step with the radio pulses.

If the core of the exploded star has a mass of more than three Suns, then even a neutron star is not the end for it. Instead, it becomes something still more bizarre: a *black hole*. No force can shore up a dead star weighing more than three solar masses against the inward pull of its own gravity. It continues to shrink, becoming ever smaller and denser until its gravity becomes so great that nothing can escape from it, not even its own light. It has dug its own grave, a black hole. Since a black hole is, by definition, invisible it is of only academic interest to amateur observers. However, professional astronomers have detected X-ray emissions from space which they believe are being given off by hot gas plunging into the bottomless pit of black holes. The best-known candidate for a black hole is Cygnus X-1; it lies near a visible 9th magnitude star in the constellation Cygnus.

Double and Multiple Stars

To the naked eye, stars appear as solitary, isolated objects. But the great majority of stars – over 75 per cent – actually have one or more companion stars, too faint or too close to be seen separately with the naked eye but which can be distinguished telescopically. Many attractive double and multiple stars are within range of binoculars and telescopes.

There are two sorts of double star. In one type, the two stars are not actually related but happen to lie in the same line of sight by chance; in this arrangement, termed an *optical double*, one star may be hundreds of light years farther from us than the other. Optical doubles are comparatively rare. Most double stars are physically linked to each other by gravity, forming a genuine *binary* system. The two stars in an optical binary orbit their mutual centre of gravity, which can take centuries. Where more than two stars are involved the orbits can become very complicated.

Twins or triplets are most common among stars, but some stellar families can be much larger. One celebrated multiple star is the so-called Double Double, ε (epsilon) Lyrae. Binoculars show it to be a wide double, but modest-sized amateur telescopes reveal that each of these stars is itself double, making a quadruple system. Even more remarkable is Castor, a system of six stars all linked by gravity. Amateur telescopes show that Castor consists of two bright blue–white stars close together, with a fainter third star some way off. Professional observations have revealed that each of these three stars is a *spectroscopic binary*. In a spectroscopic binary the two stars are too close together to be seen individually in any telescope, but analysis of the light from the star reveals the companion's presence. The two brightest stars of Castor take about four centuries to orbit each other, whereas their spectroscopic companions whirl around them in only a few days. Sometimes the members of a binary system eclipse each other as seen from Earth. These eclipsing binaries are dealt with in the next section, on variable stars.

For amateurs, the attraction of observing double stars lies in separating (or 'splitting') close doubles and in comparing their brightnesses and colours. Some stars have beautiful colour contrasts, such as the yellow and blue-green components of Albireo (β (beta) Cygni). The closer together two stars are, the larger the aperture of telescope needed to split them. If one star is much fainter than the other it will be swamped by the primary star's light, and so will be more difficult to see than a companion that is of similar brightness to the main star. The constellation notes in this book include guides to the likely apertures needed to split various double stars. But there can be no hard-and-fast rules. So much depends on the observer's eyesight, the quality of telescope, and the observing conditions. The only way to find out for sure is to look for yourself.

Variable Stars

Certain stars have a varying brightness, and are known as variable stars. Amateur astronomers can make valuable observations of them. An observer estimates the brightness of the variable star by comparing it against nearby stars of known magnitude. The observations are plotted on a graph to form what is known as a *light curve*, which can reveal much about the nature of the star under study.

The usual cause of a star's variation is actual changes in its light output, but in some cases the star is a member of a binary system in which one star periodically eclipses the other. One famous eclipsing binary star, the first of its type to be noticed, is Algol in the constellation Perseus. Algol consists of a blue dwarf, from which most of the light comes, orbited by a fainter yellow subgiant. Every 2.87 days the magnitude of Algol drops from 2.1 to 3.4 as the fainter star eclipses the brighter one; the eclipse lasts about 10 hours. Compare it at maximum with α (alpha) Persei, magnitude 1.8, and at minimum with δ (delta) Persei, magnitude 3.0. There is also a secondary minimum when the brighter star eclipses the fainter one, but the drop in light is too small for the eye to notice. Most variable star observers begin by observing Algol; regular predictions of its eclipses are issued by astronomical

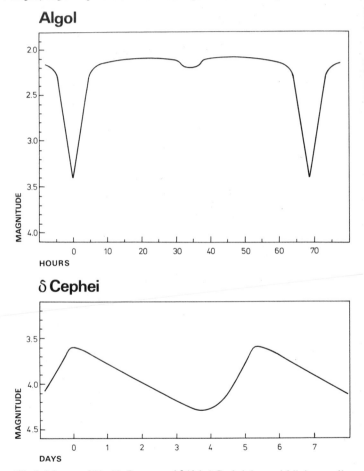

Algol

δ Cephei

The brightness of Algol in Perseus and δ (delta) Cephei rises and falls in a predictable fashion every few days, as shown by these light curves. Note that the rise of δ (delta) Cephei to maximum is quicker than its subsequent decline. The small secondary minimum of Algol is undetectable to the naked eye. *Wil Tirion.*

societies. Another eclipsing binary of note is β (beta) Lyrae, which varies between magnitudes 3.4 and 4.1 every 12.9 days. Compare it with nearby γ (gamma) Lyrae, constant at magnitude 3.2, and κ (kappa) Lyrae, magnitude 4.3.

Of the stars that vary intrinsically in brightness, most do so because of changes in their size. These are known as *pulsating variables* (not to be confused with pulsars). Of particular importance to astronomers are the so-called *Cepheid variables*, named after their prototype δ (delta) Cephei. Cepheid variables are yellow supergiant stars that go through one cycle of pulsations in periods ranging from about 2–40 days, varying in brightness by up to one magnitude as they do so. δ (delta) Cephei itself varies between magnitudes 3.6 and 4.3 every 5.4 days, and is thus an easy object for amateur observation. Good comparison stars are ε (epsilon) Cephei, magnitude 4.2, and ζ (zeta) Cephei, magnitude 3.4.

The importance of Cepheid variables is that the length of their light cycle is directly related to their absolute magnitude: the brighter the Cepheid, the longer it takes to complete one cycle of variation. Astronomers therefore have an easy way of finding these stars' absolute magnitude, simply by observing their period of fluctuation; when the absolute magnitude is compared with the apparent magnitude, the star's distance is easily computed. Cepheid variables thus make important distance indicators in astronomy.

Related pulsating stars are the RR Lyrae variables. These are old blue stars frequently found in globular clusters, and they vary by about 0.5–1.5 magnitudes in less than a day. The prototype, RR Lyrae itself, varies from magnitude 7.4–8.6 every 0.57 days. Minor additional types of pulsating variables include β (beta) Cephei stars and δ (delta) Scuti stars, both of which have short periods of only a few hours and ranges of variation too small to be noticeable to the naked eye.

Each type of variable star falls in a specific area on the Hertzsprung-Russell diagram (p. 264), and represents stars of different mass at various stages of their evolution. Variability of one kind or another seems to be an inevitable consequence of the ageing process of a star.

Red giants and red supergiants are old stars that frequently turn out to be variable. They pulsate, but not with anything like the same regularity at the types of stars mentioned above. The most abundant variables known are the long-period variables, which have periods ranging from about three months to two years, and amplitudes of several magnitudes. Their prototype is Mira (*o* (omicron) Ceti, in the constellation Cetus, the Whale), a red giant whose period averages 331 days during which it ranges from about magnitudes 3–9; the exact period and amplitude differs slightly from cycle to cycle. Another variable of the same type is χ (chi) Cygni.

More erratic are the semi-regular variables, typically with periods of about 100 days and amplitudes of 1–2 magnitudes. The irregular variables have little or no discernable pattern at all to their fluctuations. All these stars are red giants and red supergiants that have reached a stage of instability, oscillating in size and brightness. Sometimes it is difficult to decide which class a star should be placed in. Examples of semi-regular and irregular variables are Antares, Betelgeuse, α (alpha) Herculis and μ (mu) Cephei.

Most spectacular of all variable stars are the novae, which suddenly and unexpectedly erupt by perhaps 10 magnitudes or more (10,000 times or more in brightness), sometimes becoming visible to the naked eye where no star appeared before. Their name comes from the Latin meaning 'new', for they were once thought to be genuinely new stars. Now we know that they are merely old, faint stars undergoing a temporary outburst. Despite their name, they are not related to supernovae; those stars explode for different reasons.

According to current theory, novae are close double stars, one member of which is a white dwarf. Gas spilling from the companion star onto the white dwarf is thrown off in an eruption. The star does not disrupt itself in a nova outburst. In fact, some novae have undergone more than one recorded outburst, and perhaps all novae recur given time.

A nova rises to maximum brightness in a few days. Amateur astronomers are often the first to spot such eruptions and to notify professional observatories. After a few days or weeks at maximum comes a slow decline over several months, as the nova slowly sinks back to its previous obscurity, sometimes punctuated by occasional minor additional outbursts. Following the progress of a nova is one of the most fascinating aspects of variable star observation. Unfortunately, prominent naked-eye novae occur only about once every decade, but many more are visible in binoculars.

Dense Milky Way star fields, blotted by dark nebulosity, towards the centre of our Galaxy in Sagittarius. (See page 278 for Milky Way.) *Hale Observatories photograph.*

Milky Way, Galaxies and the Universe

Our Sun and all the stars visible in the night sky are members of a vast aggregation of stars known as the Galaxy (given a capital G to distinguish it from any other galaxy). Our Galaxy is spiral in shape, with arms composed of stars and nebulae that wind outwards from a central bulge of stars. It is about 100,000 light years in diameter; the Sun lies in a spiral arm 30,000 light years from the Galaxy's centre. Astronomers estimate that the Galaxy contains at least 100,000 million stars.

Most of the stars in the Galaxy lie in a disk about 2000 light years thick. Seen from our position within the Galaxy, this disk of stars appears as a faint, hazy band crossing the sky on clear, dark nights. We call the band of stars the Milky Way, and the term Milky Way is often used as a name for our entire Galaxy. The star fields of the Milky Way are particularly dense in the region of Sagittarius, which is the direction of the Galaxy's centre. Note that the plane of the Milky Way is tilted at 62° with respect to the celestial equator. This results from a combination of the tilt of the Earth's axis, and the fact that the plane of the Earth's orbit as a whole is tilted with respect to the plane of the Galaxy.

Around our Galaxy are dotted more than 100 ball-shaped clusters of stars known as *globular clusters*, several of which are visible with the naked eye or binoculars. Globular clusters contain from 100,000 to several million stars all bound together by gravity.

The brightest globular clusters are ω (omega) Centauri and 47 Tucanae, both in the southern hemisphere; in the northern hemisphere the best example is M 13 in Hercules. To the naked eye and in binoculars, these objects appear as softly glowing patches of light. Moderate-sized telescopes start to resolve some of the individual red giant stars, giving the cluster a speckled appearance. Globular clusters are believed to have formed early in the history of the Galaxy. They contain some of the most ancient stars known, 12,000 million or so years old, over twice the age of the Sun.

Our Galaxy has two small companion galaxies called the Magellanic Clouds, that appear to the naked eye like detached portions of the Milky Way in the southern constellations of Dorado and Tucana. The Large Magellanic Cloud contains about 10,000 million stars (a tenth of the number in our Galaxy) and lies about 180,000 light years from us. The Small Magellanic Cloud has about a fifth of the number of stars in the Large Magellanic Cloud, and lies somewhat farther away at 230,000 light years. Both Clouds contain numerous star clusters and bright nebulae and are rich territory for sweeping with instruments of all sizes. One can but imagine the magnificent view of our Galaxy that any astronomers living in the Magellanic Clouds would have.

Countless other galaxies are dotted like islands in the Universe as

Milky Way · star clouds and dust lanes extend northwards from Sagittarius to Scutum and Serpens. The central pink patch is M 8 (Lagoon Nebula). Two smaller patches above are M 17 (Omega Nebula) and M 16 (Eagle Nebula). *David Malin.*

far as the largest telescopes can see. Most galaxies are members of clusters containing up to thousands of galaxies. Our own Milky Way is the second-largest member of a small cluster of about 30 galaxies known as the Local Group. The largest galaxy in the Local Group is visible to the naked eye as a fuzzy, elongated patch in the constellation of Andromeda. The Andromeda Galaxy is estimated to contain about twice as many stars as are in our own Galaxy, and to be about 25 per cent greater in diameter. It lies 2.2 million light years away. Long-exposure photographs reveal that the Andromeda Galaxy is a spiral, tilted so that we see it almost edge-on. If our own Galaxy were viewed from outside, it would look like this. The Andromeda Galaxy has two small companion galaxies, its equivalent of the Magellanic Clouds, which are visible in amateur telescopes.

The only other member of the Local Group within easy range of amateur instruments is M 33 in Triangulum, another spiral galaxy, a greater distance from us than the Andromeda Galaxy and containing considerably fewer stars. It can be picked up in binoculars under clear, dark skies. The nearest rich cluster of galaxies to the Local Group lies in the constellation of Virgo, part of it spilling over into neighbouring Coma Berenices. Of its 3000 or so known members, dozens are within reach of amateur telescopes.

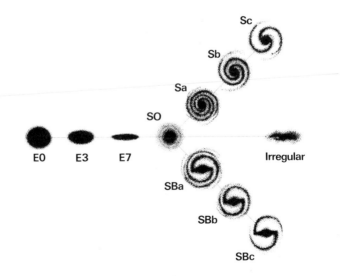

Galaxies are classified according to shape: elliptical, type E; spiral, type S; barred spiral, type SB; and irregular. *Wil Tirion.*

Astronomers classify galaxies into three main types: elliptical, spiral, and barred spiral. Elliptical galaxies range in shape from virtually spherical, known as E0, to flattened lens shapes, known as E7. They include both the largest and the smallest galaxies in the Universe. Supergiant galaxies composed of up to 10 million million stars are the most luminous galaxies known. An example is M 87 in the Virgo Cluster. At the other end of the scale, dwarf ellipticals resemble large globular clusters. Dwarf ellipticals may be the most abundant type of galaxy in the Universe, but their faintness makes them difficult to see.

Spiral galaxies (type S), such as our own Milky Way and the Andromeda Galaxy, have arms winding out from a central bulge. There are usually two arms, but sometimes more. In barred spirals (type SB), the arms emerge from the ends of a bar of stars that runs across the galaxy's centre. Spiral and barred spiral galaxies are subdivided according to how tightly their arms are wound: types Sa and SBa have the tightest-wound arms while type Sc and SBc have the loosest-wound arms. Our own Galaxy falls midway between Sb and Sc; M 31 in Andromeda is type Sb.

Most of the galaxies observed in the Universe are large, bright spirals, but as mentioned above they may actually be outnumbered by dwarf galaxies that are too faint to see over great distances. In addition to these three main types, there are certain galaxies classified as irregular; the Magellanic Clouds fall into this class, although there is some semblance of spiral structure in the Large Magellanic Cloud.

Being faint, fuzzy objects, galaxies are best seen where skies are clear and dark, away from city haze and stray light. Low magnification is best, to increase the contrast of the galaxy against the sky background. Through a telescope you should see the nucleus of a galaxy as a starlike point surrounded by the misty halo of the rest of the galaxy. Do not expect to see spiral arms as on a long-exposure photograph. Galaxies present themselves to us at all angles, so even spirals can seem to be elliptical in shape when viewed edge-on.

Peculiar things are going on within some galaxies. For example, certain galaxies give off intense amounts of energy as radio waves; these are called *radio galaxies*, and include the supergiant ellipticals M 87 in Virgo and NGC 5128 in Centaurus. Some spiral galaxies have unusually bright nuclei. These are termed *Seyfert galaxies*, after the astronomer Carl Seyfert who drew attention to them in 1943. M 77 in Cetus is the brightest Seyfert galaxy.

Most peculiar of all are the objects known as *quasars*, which emit as much energy as hundreds of normal galaxies from an area less than a light year in diameter. Despite their exceptional nature quasars are not at all exciting visually, which is why they were overlooked until 1963. The brightest quasar, 3C 273, appears as an unimposing 13th magnitude star in Virgo. Most astronomers now believe that quasars are actually

The beautiful Whirlpool Galaxy, M 51, in Canes Venatici. At the end of an arm lies the irregular companion galaxy NGC 5195. Photographed with 3.6 m Canada-France-Hawaii Telescope by Laird Thompson. *University of Hawaii.*

A dark lane of dust rings the star-packed core of the giant elliptical galaxy NGC 5128 in the constellation Centaurus. This odd galaxy is also known as the radio source Centaurus A. *Anglo–Australian Telescope Board.*

the active centres of galaxies so far off in the Universe that their o
regions are too faint to be seen. Radio galaxies, Seyferts and qua.
are probably all related in some way; perhaps they are all you
galaxies seen at different stages of their early evolution. One favour
theory is that their central powerhouse is a massive black hole th.
gobbles up stars and gas from the surrounding Galaxy.

In 1929 the American astronomer Edwin Hubble discovered that
the galaxies are moving apart from each other as though the Universe
is expanding like a balloon being inflated. (But clusters of galaxies such
as the Local Group are not expanding – they are held together by their
mutual gravitational attraction.) Hubble's discovery that the Universe
is expanding came from a study of the spectrum of each galaxy's light.
This revealed that the light from the galaxies was being lengthened in
wavelength as a result of high-speed recession (which is called the
Doppler effect). Such a lengthening of wavelength is called a *red shift*,
because the light from the galaxy is moved towards the red (longer
wavelength) end of the spectrum. Incidentally, the red shift does not
make galaxies actually look redder, because the blue end of the spectrum is filled in by light that was formerly at ultraviolet wavelengths.

Hubble found that the amount of red shift in a galaxy's light is
directly related to its distance, the most distant galaxies having the

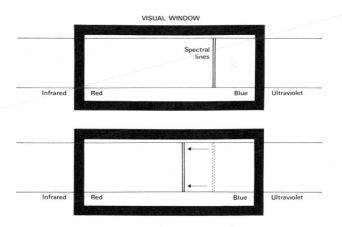

The red shift in the light from a distant galaxy is measured by the change in
position of lines in its spectrum. The change in position of the lines is here shown
with reference to the visual window, the range of wavelengths to which the
human eye is sensitive. All the light from the galaxy is shifted by the same amount,
so while some of the galaxy's light moves out of sight beyond the red end of the
visual window, other light moves into the visible region from the ultraviolet.
Wil Tirion.

est red shifts. Therefore by measuring the red shift of a galaxy nomers can tell how far away it is. Quasars, for instance, exhibit enormous red shifts that they must be the most distant objects ble in the Universe, up to 15,000 million light years away.

Since the Universe is expanding, it is logical to conclude that it was nce smaller and more densely packed than it is now. According to the most widely accepted theory, the entire Universe was originally a compressed superdense blob which, for some unknown reason, exploded in a cataclysm known as the Big Bang. The galaxies are the fragments from that explosion, still flying outwards. As far as anyone can tell, the Universe will continue to expand for ever. According to the best estimates the Big Bang took place between 10,000 million and 20,000 million years ago; that is the age of the Universe as we know it. It is impossible to tell what, if anything, happened before the Big Bang.

NGC 4151 in Canes Venatici is an example of a Seyfert galaxy, a type of spiral galaxy with a bright but variable centre. Seyfert galaxies are thought to be closely related to quasars. *Hale Observatories photograph.*

285

The Sun

Our Sun is a glowing sphere of hydrogen and helium gas 1.4 milli
km in diameter, 109 times the diameter of the Earth and 745 times
massive as all the planets put together. The Sun is vital to all life o
Earth, because it provides the heat and light without which our plane
would be uninhabitable. It is important to astronomers because it is
the only star that we can observe in close-up. Studying the Sun tells
us a lot that we could never otherwise know about stars in general.

Whereas most celestial objects present problems for observers because they are so faint, the Sun provides us with entirely the opposite
problem: it is so bright that it is dangerous to look at. Anyone glancing
for an instant at the Sun through any form of optical instrument, be it
binoculars or telescope, risks instant blindness. Even staring with the
naked eye at the Sun for long periods can permanently damage your
eyesight. There is only one safe way to observe the Sun, and that is by
projecting its image onto a piece of white card. Some telescopes come

The Sun on July 4, 1974, at a time of relatively low solar activity. The large
sunspot group could engulf the Earth several times over. Note the effect of limb
darkening, against which brighter faculae may be seen. *Kitt Peak National
Observatory.*

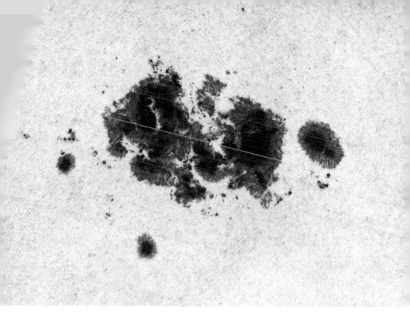

Large group of sunspots 200,000 km long (as it appeared on May 17, 1951). Filamentary structure in the outer penumbra of the spots gives them a sunflower-like appearance. Note also the granulated appearance of the surrounding solar photosphere. *Mount Wilson Observatory photograph.*

equipped with dark Sun filters to place over the eyepiece. These should never be used, for they can crack during observation, with inevitably disastrous results.

A view of the projected image of the Sun displays the brilliant solar surface, or photosphere ('sphere of light'), consisting of seething gases at a temperature of 5500°C. Although this is intensely hot by terrestrial standards, it is cool by comparison with the Sun's core where the energy-generating nuclear reactions take place; there, the temperature is calculated to be about 15 million °C.

The photosphere exhibits a mottled, rice-grain effect termed *granulation*, caused by cells of hot gas bubbling up in the photosphere like water boiling in a pan. Granules range from about 300 km to 1500 km in diameter. Look carefully at the projected image of the Sun and you will notice that the edges of the Sun appear fainter than the centre of the disk, an effect known as *limb darkening*. This is caused by the fact that the gases of the photosphere are somewhat transparent, so that at the centre of the disk we are looking more deeply into the Sun's interior than at the limbs. Brighter patches known as *faculae* may be visible against the limb darkening. Faculae are areas of higher temperature on the photosphere. A number of prominent dark markings known as *sunspots* will probably also be visible. They are areas of cooler gas t appear dark by contrast against the brighter photosphere.

Sunspots are temporary features that occur where magnetic fields c the Sun burst through the photosphere. Apparently, the presence of strong magnetic field blocks the outward flow of heat from inside the Sun, causing the cooler spot. Sunspots have a dark centre known as the umbra, of temperature about 4000°C, surrounded by a brighter pen-umbra at about 5000°C. (These temperatures, incidentally, are about the same as on the surface of a red giant such as Aldebaran.)

Sunspots range in size from small pores no bigger than a large granule to enormous, complex patches 100,000 km or so across. The largest sunspots tend to form in groups that could span the distance from the Earth to the Moon. Spots that big are visible to the naked eye when the Sun is dimmed by the atmosphere shortly before sunset. A major sunspot takes about a week to develop to full size, then slowly dies away again over the next fortnight or so.

As the Sun rotates, new spots are brought into view at one limb while old groups disappear around the other limb. Spot groups seldom last more than one or two solar rotations (a month or two). Usually, a spot group contains two main components oriented east–west. The spot that leads across the disk as the Sun rotates is termed the p (preceding) spot, and is usually larger than its follower, the f spot. The p and f spots have opposite magnetic polarities, like the ends of a horseshoe magnet. The horseshoe is completed by invisible lines of force looping between the spots.

Sometimes the intense magnetic fields in a complex sunspot group become entangled, releasing a sudden flash of energy known as a flare that may last from a few minutes to an hour. Atomic particles are spewed out into space by the eruption of a flare. These particles reach Earth about a day later, causing effects in the upper atmosphere such as radio interference and the glowing displays known as *aurorae*. Un-fortunately, aurorae occur only near the Earth's magnetic poles, so you will not be able to see them often unless you live in far northerly or southerly latitudes. An auroral display is an ethereal sight: the sky seems to glow with coloured light that can take the form of folded drapery or arches, shimmering and changing shape for hours on end.

By observing the progress of spots as they are carried across the Sun's face we can measure the rotation period of the Sun. Being gaseous and not solid, the Sun does not rotate at the same rate at all latitudes. It spins most quickly at its equator, once every 25 days; this falls to about once every 28 days at latitude 45°, and near the poles it is slowest of all, once every 34 days. The average figure usually quoted, 25.38 days, refers to the rotation rate at latitude 15°. As a result of the Earth's orbital motion around the Sun, a spot takes about two days extra to return to the same place on the Sun as seen from Earth.

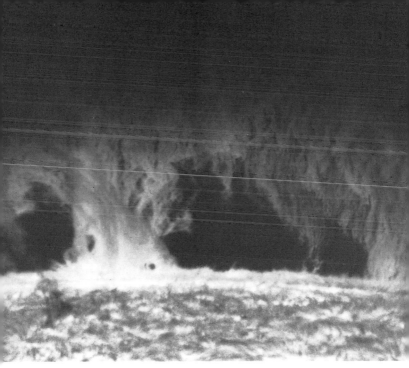

Solar prominences extending 65,000 km into space look like enormous burning trees at the edge of the Sun. Photograph taken in the red light of hydrogen. *Big Bear Solar Observatory*.

The number of sunspots on view waxes and wanes in a cycle lasting 11 years on average, although the length of individual cycles has been as short as 8 years and as long as 16 years. At times of minimum activity the Sun may be spotless for days on end, whereas at solar maximum over 100 spots may be visible at any one time. However, the level of activity varies considerably from cycle to cycle; the average number of sunspots visible at maximum has ranged from 40–180, and even when the Sun is supposedly quiet some large spots and flares can break out. Solar activity is notoriously unpredictable, which adds to its fascination for observers.

A few general rules can be deduced, however. The first spots of each new cycle appear at latitudes of about 30–35° north and south of the Sun's equator. As the cycle progresses, spots tend to form closer to the equator. Sunspot numbers build up to a peak, then start to decay again. As minimum approaches, the last spots of the cycle are found at latitudes between 10° and 5° north and south of the equator. At the same time – that is, around solar minimum – the first spots of the next cycle are forming at higher latitudes. Sunspots are seldom found on the Sun's equator, and spots at latitudes greater than 40° are exceedingly rare.

Above the photosphere is a tenuous layer of gas known as the *chromosphere*, about 10,000 km deep. So faint is the chromosphere that it is normally invisible except with special instruments. It can, however, be seen for a few seconds at a total eclipse when it appears as a pinkish crescent just before and after the Moon completely covers the face of the Sun. Its pinkish colour is caused by light from hydrogen gas, and gives rise to its name, which means 'colour sphere'.

Also visible around the edge of the Sun at total eclipses are huge clouds of gas extending from the chromosphere into space, known as *prominences*. They have the same characteristic rosy-pink colour as the chromosphere, caused by emission from hydrogen. Like so many features of the Sun, prominences are controlled by magnetic fields. So-called quiescent prominences extend for 100,000 km or more across the Sun, sometimes forming graceful arches tens of thousands of kilometres high. When seen silhouetted against the brighter background of the photosphere, they are termed filaments. Quiescent prominences can last for months. At the other end of the scale are eruptive prominences with lifetimes of only a few hours. They are flares seen at the edge of the Sun, ejecting material into space at speeds of up to 1000 km per second. All forms of solar activity – sunspots, flares and prominences – follow the 11-year solar cycle.

The Sun's crowning glory is its *corona*, a faint halo of gas that comes into view only when the brilliant photosphere is blotted out at a total solar eclipse. The corona is composed of exceptionally rarefied gas at a temperature of 1–2 million °C. Petal-like streamers of coronal gas extend from the Sun's equatorial zone, while shorter, more delicate plumes fan out from the polar regions. The shape of the corona changes during the solar cycle; at solar maximum, when there are more active areas on the Sun, the corona appears more regular in shape than at solar minimum.

Gas from the corona is continually flowing away from the Sun and into the Solar System, forming what is known as the *solar wind*. Atomic particles of the solar wind are detected streaming past the Earth at a speed of about 400 km per second. The most obvious effect of the solar wind is to make comet tails point away from the Sun. The solar wind extends outwards beyond the orbit of the most distant planet, finally merging with the thin gas between the stars. In a sense, therefore, all the planets of the Solar System can be said to lie within the outer reaches of the Sun's corona.

Corona of the Sun, photographed from the ısland of Zanzibar at the total solar eclipse of October 23, 1976. *Paul Cass*.

The Solar System

In one sense our Sun is unusual, for it is not accompanied by another star but has a family of nine planets, assorted moons and countless smaller lumps of debris. (Any planets around other stars are too faint to be seen, at least from optical telescopes on Earth.) This retine of bodies, all held captive by the Sun's gravity, is called the Solar System.

All the objects in the Solar System shine by reflecting the light coming from the Sun. Several of the planets when at their brightest can equal or outshine the brightest stars. All the planets orbit the Sun in approximately the same flat plane, so they are always to be found near the ecliptic. A bright 'star' that disturbs a familiar constellation shape along the ecliptic is therefore almost certainly a planet (although it could, just possibly, be a nova).

As seen from above the Sun's north pole, the planets orbit the Sun in an anticlockwise direction. Their orbits are slightly elliptical in shape, so that each planet's distance from the Sun varies somewhat during one orbit. In order of average distance from the Sun, the nine planets are: Mercury, Venus, Earth, Mars, Jupiter, Saturn, Uranus, Neptune and Pluto. The four inner planets are rocky, relatively small bodies. Then come four giant planets made mostly of gas. Pluto is a small, frozen oddity on the outer limits.

Venus is the planet with the most circular orbit; its distance from the Sun varies by little more than 1.5 million km. The planet with the most elliptical orbit is distant Pluto, which at times actually strays across the orbit of Neptune. Some of the minor bodies, or asteroids, in the Solar System have orbits that are more elliptical than this, and the comets have highly distended orbits that can take them from the innermost part of the Solar System out to way beyond Pluto. The closest point of a celestial body's orbit to the Sun is known as the *perihelion*; the farthest point is the *aphelion*.

The basic unit of distance in the Solar System is the *astronomical unit*, the average distance of the Earth from the Sun; it is equivalent to 149,597,870 km. Light takes 499 seconds (8.3 minutes) to cross this distance, so we on Earth see the Sun as it actually appeared over 8 minutes ago. Clearly, although the astronomical unit is large compared with everyday distances on Earth, it is insignificant in comparison with a light year. There are 63,240 astronomical units in a light year.

How long a planet takes to orbit the Sun depends on its distance: the closest planets orbit the most quickly, and the farthest orbit the slowest. Each planet's orbital period is technically its 'year', although the period is usually expressed in terms of Earth days and years. The orbital

Above: The sizes of the planets to scale, compared with a section of the Sun.

Below: The planets of the Solar System in order of distance from the Sun. *Wil Tirion.*

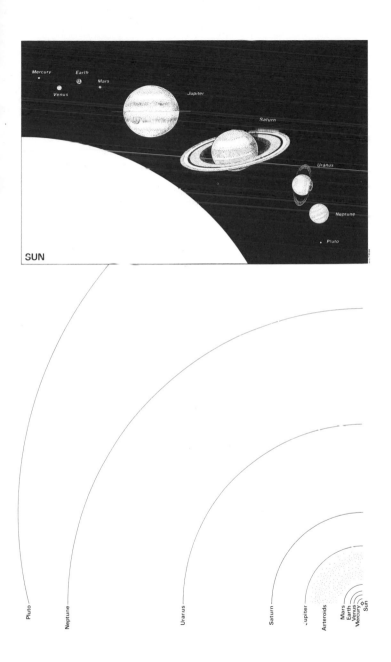

SUN

Mercury

Earth

Venus

Mars

Jupiter

Saturn

Uranus

Neptune

Pluto

Pluto

Neptune

Uranus

Saturn

Jupiter

Asteroids

Mars

Earth

Venus

Mercury

Sun

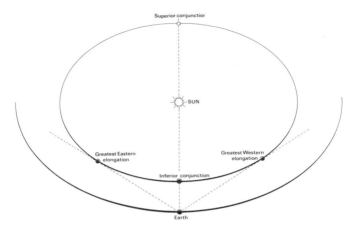

When Venus or Mercury lie directly between the Earth and the Sun they are said to be at inferior conjunction; when on the far side of the Sun they are at superior conjunction. Their maximum angular separation from the Sun is called greatest elongation. *Wil Tirion.*

periods of the planets range from 88 days for Mercury to 250 years for Pluto. These orbital periods are known technically as *sidereal periods*, for they are measured with respect to the distant stars.

From time to time, as Mercury and Venus move around their orbits, they come between us and the Sun; when that happens, they are said to be at *inferior conjunction.* That is also the time when Mercury and Venus are closest to the Earth, although this fact is of little use to observers since the planets are lost in the Sun's glare and their illuminated hemispheres are in any case turned away from the Earth.

The orbits of Mercury and Venus are inclined by 7° and 3.4° respectively to the orbit of the Earth, but that is sufficient to ensure that they usually pass above or below the disk of the Sun at inferior conjunction. But occasionally Mercury or Venus do actually cross the face of the Sun as seen from Earth. At such an event, called a transit, the planet appears as a tiny dark dot like a small sunspot against the glaring solar surface. Transits of Mercury are more common than those of Venus: they are due in 1986, 1993, 2003, 2006 and 2016, whereas the transits of Venus in 2004 and 2012 will be the only ones for over a century.

When Mercury and Venus lie on the far side of the Sun from us they are said to be at *superior conjunction.* When Mars and the planets beyond it lie on the far side of the Sun they are simply said to be at conjunction – being farther away from the Sun than we are they cannot possibly come between Earth and the Sun, so there is no ambiguity about which sort of conjunction is meant. Planets at conjunction are invisible in the Sun's glare.

The best time to view the two inner planets is when they are at their maximum possible angular separation from the Sun, known as *greatest elongation*. At greatest elongation, Mercury and Venus are at exactly half phase. An eastern elongation means that the planets are setting after the Sun in the evening sky; at western elongation, the planets rise before the Sun in the morning sky. The time between one greatest eastern elongation of Mercury and the next, or one greatest western elongation and the next, is 116 days. In the case of Venus, greatest western elongations or greatest eastern elongations recur every 584 days, although Venus is so prominent that it can be adequately observed well away from greatest elongation. Incidentally, the time between one given appearance of a planet and the next – be it conjunction, elongation or whatever – is known as the *synodic period*. This differs from the sidereal period because our observation platform, the Earth, is moving in orbit around the Sun.

The best opportunities to see Mars and the outer planets is when they are directly opposite the Sun in the sky; this is termed *opposition*. A planet at opposition appears due south for northern hemisphere observers (due north for observers in the southern hemisphere) at midnight local time, or 1 a.m. in summer time. At opposition Mars, Jupiter, Saturn, Uranus, Neptune and Pluto lie closest to Earth (Mercury and Venus, being between the Earth and the Sun, cannot come to opposition). These outer planets therefore appear biggest and brightest around the time of opposition.

Mars, Venus and Jupiter as seen from Earth. On Mars the dark wedge of Syrtis Major is visible, as is the southern polar cap. The photograph of Venus was taken in ultraviolet light to bring out the cloud features. *Lowell Observatory*.

The Moon

The Moon, the Earth's natural satellite and nearest celestial neighbour, is an object of perennial fascination for observation with instruments of all sizes. Despite its small size – 3476 km in diameter, roughly a quarter that of the Earth – it is so close, on average 384,400 km, that even ordinary binoculars reveal a wealth of detail on its cratered surface. Some of the most interesting objects to look out for are shown in the maps on pages 316 to 327.

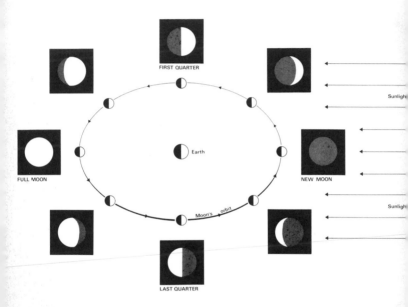

The phases of the Moon occur as the Moon orbits the Earth every month; we thus see differing proportions of its illuminated side. *Wil Tirion*.

Every month the Moon goes through a cycle of phases from new (un-illuminated) through waxing half (first quarter), full, waning half (last quarter), back to new again. Strictly speaking, there are two types of month. The first lasts 27.3 days, the time that the Moon takes to complete one orbit of the Earth relative to a fixed point such as the distant stars; this is known as a *sidereal month*. But the Earth also moves relative to the Sun in that time, so the Moon must complete rather more than one orbit to return to the same phase as seen from Earth. The time that

The Moon at gibbous phase, 10 days after new. Note Sinus Iridum on the terminator at top. Most of the features on this photograph are visible in small telescopes or even binoculars, if mounted firmly. *Lick Observatory.*

the Moon takes to complete one cycle of phases, 29.5 days, is called a *synodic month*. Each spot on the Moon is subjected to two weeks of daylight, during which surface temperatures reach the boiling point of water (about 100°C), followed by a two week night when temperatures plummet to −170°C.

The Moon spins on its axis in 27.3 days, the same time that it takes to complete one orbit of the Earth – what is known as a captured rotation – so that the same face of the Moon is always turned towards us. In practice, though, we can see slightly more than half the Moon's surface. The Moon's orbit is inclined at about 5° to the ecliptic and the Moon's axis is tilted about 1.5° relative to its orbit, so that at times we can see as much as 6.5° over the Moon's north or south poles; this is known as *libration in latitude*. In addition, the Moon's speed of motion along its elliptical orbit changes rhythmically as it approaches and recedes from the Earth, while its axial rotation remains uniform. The Moon therefore seems to rock slightly to west and east as it orbits the Earth, so that we can see up to 7° or so around each edge; this is known as *libration in longitude*. The net effect of these librations is that we can see 59 per cent of the Moon's surface.

The line dividing the lit and unlit portions of the Moon is known as the terminator. Objects near the terminator are thrown into sharp relief by the low angle of illumination, so that craters and mountains appear particularly rugged. As the Sun rises higher over the Moonscape details become more washed out. Near full Moon many individual formations are difficult to pick out. An exception are those craters that have bright ray systems, apparently made of pulverized rock thrown out from the crater during its formation; the rays become more prominent under high illumination. It is easiest to pick out the difference in contrast between the bright highlands and the dark lowland plains known as maria (singular: mare) at full Moon.

After looking at the brilliant full Moon it comes as a surprise to realize that the lunar surface rocks are actually dark grey in colour; on average, the Moon's surface reflects only seven per cent of the light that hits it. If the Moon were, for instance, covered in clouds like those of Venus it would be over 10 times brighter.

Formations on the Moon bear a variety of curious names. The dark lowland plains are known as *maria*, Latin for seas, because the first observers imagined them to be stretches of water; the name persists, even though it has been clear for centuries that the Moon is both airless and waterless. Thus we have for instance the Oceanus Procellarum (Ocean of Storms), Mare Imbrium (Sea of Rains) and Mare Tranquillitatis (Sea of Tranquillity). Less prominent lowlands are characterized as bays (Sinus), marshes (Palus) or lakes (Lacus). Mountains on the Moon are named after terrestrial ranges; hence we have the lunar Alps and the lunar Apennines. The craters have been named after

Apollo 17 astronaut Harrison Schmitt uses a rake to scoop up rocks from the dark and dusty plain of the Moon's southeastern Mare Serenitatis, contrasting with the bright foothills of the Taurus Mountains in the background. *NASA*.

philosophers and scientists of the past, though it is fair to say that the selection has been somewhat arbitrary.

We owe the modern system of lunar nomenclature to an Italian astronomer, Giovanni Riccioli, who in 1651 published a map containing many of the names now universally accepted. Riccioli allocated himself a prominent crater near one edge of the Moon, with a large neighbouring formation named after his pupil, Francesco Grimaldi. Less favoured colleagues did not fare so well. Galileo, one of the greatest scientists of all time, but who fell out with the Church, is given an insignificant 15 km diameter crater in Oceanus Procellarum.

Riccioli's system has been followed to the present day, but now that the near and far sides of the Moon have been mapped in detail by spacecraft, the ground rules have been extended to include names from areas of human endeavour other than astronomy. Thus we now find Freud, H. G. Wells and Montgolfier commemorated on the Moon. Finding suitable names for features on celestial bodies has become a pressing problem now that many planets and their moons have been surveyed by space probe.

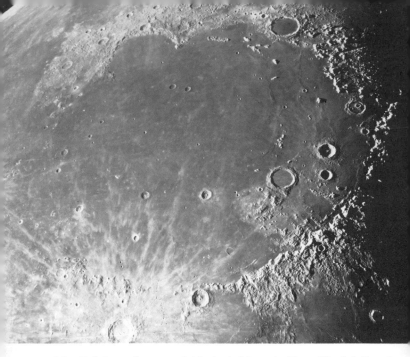

Mare Imbrium: a large, rounded lowland plain on the Moon. The dark-floored crater Plato pierces the mountains on its northern shore. The Mare's southern reaches are splashed with Copernicus' bright rays. *Mount Wilson Observatory*.

Lunar observation began in 1609, when Galileo turned his first telescope towards the Moon. His drawings, published in 1610, are crude by modern standards but served to show that the Moon's surface is rugged and mountainous. Debate raged for the next three and a half centuries over the origin of the Moon's features. One school of thought held that they were caused by volcanic action; the opposition believed the craters and maria to have been blasted out by massive impacts with meteorites and asteroids. There is no point in recapitulating the arguments here. Suffice it to say that the impact theory has emerged as the undisputed winner, although it is accepted that there has also been a certain amount of volcanic activity on the Moon.

Not until the late 1960s, when the first space probes and the first men reached the Moon, was the controversy resolved. On July 31, 1964, an American probe called Ranger 7 headed towards a corner of the Mare Nubium. As it zoomed close in it sent back a stream of photographs showing features as small as a metre across, a hundred times better than what is visible through Earth-based telescopes. Rangers 8 and 9 followed the next year. The lesson from these probes was that even the apparently flattest parts of the Moon's surface were pitted with small craters caused by aeons of meteorite bombardment. Landing sites for future manned spacecraft would therefore have to be chosen with particular care, to prevent the landing craft from toppling into a crater or hitting a boulder (this nearly happened on the Apollo 11 descent).

Ranger's cursory examination was followed by a two-pronged attack: Surveyor, a series of craft which soft-landed automatically to give an astronaut's eye view of the lunar surface (the Rangers had simply crashed), and Lunar Orbiter which, as the name implies, photographed the Moon from close orbit. Together, these two series of probes revolutionized our knowledge of the Moon between 1966–68, and paved the

Oblique view of the twin craters Aristarchus (bright, with rays) and Herodotus, taken with a mapping camera aboard an Apollo spacecraft. From a small crater near Herodotus emerges the snake-like Schröter's Valley. *Fairchild*.

Apollo 17 astronaut Harrison Schmitt is dwarfed by a massive boulder during his exploration of the Moon's surface in December 1972. *NASA*.

way for the manned Apollo landings. The Surveyors showed that the Moon's topsoil is made of compacted dust, firm enough to support astronauts and their spacecraft. From Lunar Orbiter photographs, astronomers compiled their most detailed maps of the near and far sides of the Moon.

The Moon's far side was first glimpsed by the Soviet probe Luna 3 in October 1959. Its photographs were poor by modern standards, but they did at least reveal the main difference between the two hemispheres of the Moon: there are scarcely any mare areas on the Moon's far side. Instead, heavily cratered bright uplands dominate the scene. The reason for this asymmetry is that the Moon's crust is several kilometres thicker on the far side. Large lowland basins like Mare Imbrium exist on the far side of the Moon, but they have not been flooded by dark lava. Volcanic lavas from the Moon's interior found it easier to leak out through the thinner crust of the Earth-facing hemisphere. The most prominent dark area on the Moon's far side is not a true mare at all, but a deep crater called Tsiolkovsky, 240 km in diameter (almost as large as Sinus Iridum on the visible hemisphere).

Once the Lunar Orbiters had spied out potential landing sites, the stage was set for men to follow. On July 20, 1969, the Apollo 11 lunar module called Eagle carried Neil Armstrong and Edwin Aldrin to the first manned lunar landing, in the southwestern Sea of Tranquillity. The two astronauts spent two hours exploring the Moon's surface, setting up experiments and collecting samples for geologists to study back on Earth. For the first time, humans had touched another world in space.

By the time Apollo 17 concluded the series of manned landings in December 1972, astronauts had brought back to Earth over 380 kg of Moon samples, most of which is stored in Houston, Texas. When divided into the overall cost of Apollo, one kilogram of Moon rocks is worth 100 million U.S. dollars. In addition to the Apollo samples, three Soviet automatic lunar probes have returned to Earth with a few hundred grams of Moonsoil.

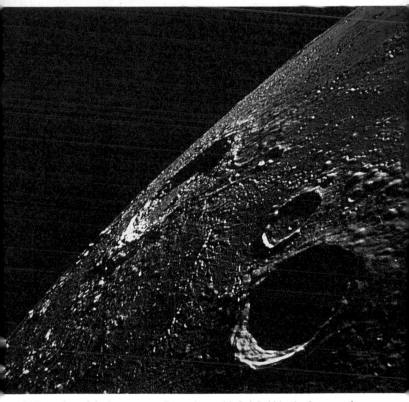

Oblique view of the lunar crater Copernicus, with Reinhold in the foreground. The surface of Oceanus Procellarum is splashed with ejecta thrown out by the meteorite impact that formed Copernicus about 1000 million years ago. *NASA*.

What have we learned from these precious specimens? The most astounding fact about the Moon rocks was their immense age. The Apollo 11 samples, for instance, proved to be 3700 million years old, older than virtually any rocks on Earth – yet the site from which they came, Mare Tranquillitatis, is one of the youngest areas on the Moon. The youngest Moon rocks of all, found at the Apollo 12 site in Oceanus Procellarum, are 3200 million years old. As expected, the lunar maria turned out to be covered with lava flows similar in composition to volcanic basalt on Earth. They do not resemble a jagged, lumpy lava field on Earth because over the thousands of millions of years since the lavas were laid down, the sandblasting effect of micrometeorites has eroded the surface rocks to form a layer of soil several metres deep, called the *regolith*.

By contrast the highlands, sampled by later Apollo missions, consist of a paler rock called anorthosite, rare on Earth. Rocks from the highlands proved older than those from the maria, mostly dating back 4000 million years or earlier. Their jumbled, fragmented nature bore witness to the violent bombardment from meteorites which the Moon suffered early in its history.

Despite scientists' hopes, the rich haul of Moon rocks did not conclusively answer the question of the Moon's origin. Differences in chemical composition between Moon and Earth seem to rule out the theory that the Moon is a fragment of our planet, but there is still no agreement on whether the Earth and Moon formed side by side, much as they are now, or whether the Moon was once a separate body that was subsequently captured by the Earth's gravity.

That problem aside, we now have a much clearer picture of the rest of the Moon's history. Evidence from the rocks reveals that the Moon formed 4600 million years ago, at the same time as the Earth and, presumably, the rest of the Solar System. Each body in the Solar System is believed to have been built up by an accumulation of many smaller bodies. This process, termed *accretion*, released sufficient heat to melt the Moon. A scum of less dense rock formed a primitive crust which was battered by the infall of debris left over from the formation of the Solar System. This bombardment carved out the Moon's cratered highlands and the mare basins. Other rocky bodies in the Solar System, notably Mercury, also bear the scars of this same mopping-up operation.

About 4000 million years ago the storm of debris abated. Then, slowly, molten lava began to seep out from inside the Moon, solidifying to form the dark lowland maria. Wrinkles of solidified lava are visible in binoculars and small telescopes when the maria are under low illumination. In particular, look out for the Serpentine Ridge in eastern Mare Serenitatis when the Moon is five to six days old, or six days past full. Mare Tranquillitatis and Mare Imbrium also have prominent wrinkle ridges. In fact, one 'crater' on Mare Tranquillitatis, called Lamont,

Meandering near the foothills of the Apennine Mountains, Hadley Rille is a dried-up lava channel that was visited by the Apollo 15 astronauts in 1971. *NASA.*

consists of nothing more than low ridges of solidified lava. In places, particularly in western Oceanus Procellarum, a number of blister-like domes have been produced by the upwelling of molten lava. But volcanic cones like Vesuvius are absent on the Moon, evidently because the lunar lavas were too runny to build up into mountains.

By 1000 million years ago the volcanic outpourings had ceased, leaving the Moon cold and dead. Since then it has remained virtually unchanged, save for the arrival of an occasional meteorite to punch a new crater in its surface. For instance, the crater Copernicus was formed roughly 1000 million years ago; Tycho was blasted out about 300 million years ago.

Yet the Moon may not be completely inactive even today. Observers have from time to time reported transient events, such as glows and obscurations, around the edges of the maria and in certain craters – Aristarchus seems to be a favourite spot for these transient lunar phenomena (TLPs). Most TLPs have been observed by amateur astronomers. Their origin remains puzzling, but they are probably

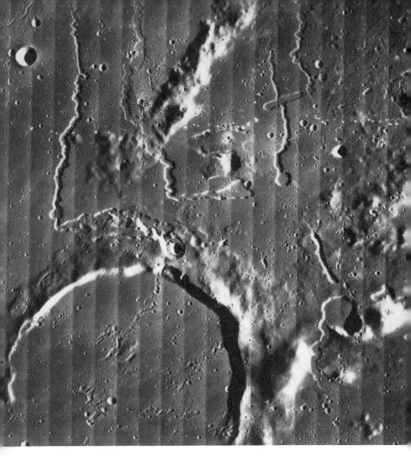

Rilles and faults on the surface of Oceanus Procellarum to the north of the horseshoe-shaped crater Prinz, photographed by Lunar Orbiter V. The picture appears banded because it was transmitted to Earth in strips. *NASA*.

caused by the release of gas from the Moon's interior.

Other features to look out for when observing the Moon are grooves in the surface, known as *rilles*, apparently caused by faulting. For instance, the crater Hyginus near the Moon's centre lies in the middle of a long, rimless cleft along which smaller craters have been formed by subsidence. Another type of rille, called a sinuous rille, snakes over the mare surface like a meandering river. Apollo 15 landed at the edge of one of these sinuous rilles, called Hadley Rille. Such winding rilles are not dried-up river beds but are probably collapsed tunnels through which underground lava streams once flowed.

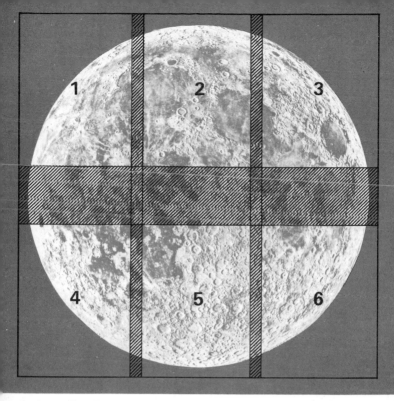

The image shows the Moon divided into six numbered sections (1, 2, 3 across the top row and 4, 5, 6 across the bottom row).

A note on orientation: These maps show the Moon in conventional orientation, with north at the top, as it appears to the naked eye and in binoculars. Through an astronomical telescope, south will be at the top, so telescope users must invert the maps. West is at the left, as on Earth, although in the rest of the sky west is in the opposite direction, towards the western horizon.

Moon Map 1

Aristarchus Brilliant young crater 45 km in diameter with multiple terracing on its inner walls. The brightest area on the Moon, and centre of a major ray system. Dark bands are visible on the inner walls under high illumination. Aristarchus is the location of numerous red glows known as transient lunar phenomena (TLPs), apparently due to outgassing from the surface. Mountains to the north exhibit fantastic sculpturing.

Copernicus Diameter 93 km. One of the most magnificent craters on the Moon. Centre of a major ray system. Terraced walls and numerous hummocky central peaks. Rays from Copernicus are splashed for more than 600 km across the Mare Imbrium and Oceanus Procellarum.

Encke Low-walled crater 29 km in diameter covered by rays from nearby Kepler. Brilliant under high illumination.

Euler Small, sharp crater, 28 km across, in southwestern Mare Imbrium. Centre of a minor ray system.

Fauth Keyhole-shaped crater south of Copernicus, nearly 2 km deep. Striking under low illumination.

Harpalus Crater 40 km in diameter on Mare Frigoris, bright under high illumination. The smaller crater Foucault, 24 km in diameter, lies between it and Sinus Iridum.

Herodotus Companion to Aristarchus, similar in size (diameter 35 km) but different in structure – it has a dark, lava-flooded floor and it is not a ray centre. The W-shaped Schröter's Valley, 200 km long, starts at a craterlet outside the northern wall of Herodotus.

Hevelius Large, bright crater 118 km in diameter on western shore of Oceanus Procellarum, with clefts crossing the floor. Smaller, sharper crater Cavalerius (64 km diameter) adjoins to the north.

Kepler Major ray centre in Oceanus Procellarum. Brilliant crater, 32 km in diameter, with central peak and heavily terraced walls.

Mairan Prominent crater in highlands west of Sinus Iridum. Diameter 41 km.

Marius Dark, flat-floored ring 41 km in diameter on Oceanus Procellarum, notable because it lies on a wrinkle ridge. There are many dome-like structures in this area where lava has bubbled up through the surface.

Oceanus Procellarum Vast dark plain with no clear-cut borders extending from Mare Imbrium southwards to Mare Humorum. Oceanus Procellarum occupies an area of more than two million km^2. It is dotted with numerous craters and bright rays.

Prinz U-shaped formation 52 km in diameter, remains of a half-destroyed crater flooded by lava from Oceanus Procellarum. Notable because of sinuous rilles to the north.

Pythagoras Magnificent large crater 128 km in diameter with terraced walls rising nearly 5 km and prominent central peak, near northwestern limb of the Moon.

Reiner Sharp crater 30 km in diameter west of Kepler in Oceanus Procellarum. Particularly notable is a tadpole-shaped splash of brighter material on the dark plain to the west and north.

Reinhold Prominent crater 48 km in diameter southwest of Copernicus, with terraced walls. A smaller, lower ring to the northeast is Reinhold B.

Rümker Remarkable formation on northwestern Oceanus Procellarum, visible only under low illumination. Rümker is an irregular, lumpy dome 55 km wide.

Sinus Iridum Beautiful large (260 km in diameter) bay on the Mare Imbrium. Its seaward wall has been broken down by invading lava and is reduced to a few low wrinkle ridges. Its remaining walls, angular in shape, are known as the Jura Mountains, which are brilliantly illuminated by the morning Sun. A deep, 39 km diameter crater on the northern rim is Bianchini.

Moon Map 2

Agrippa Oval-shaped crater 46 km in diameter with central peak, forming a neat pair with Godin.

Alpine Valley Flat floored valley 150 km long through the lunar Alps, connecting Mare Imbrium and Mare Frigoris.

Anaxagoras Crater 51 km in diameter near the north pole of the Moon, centre of an extensive ray system.

Archimedes Distinctive, flooded ring 83 km in diameter in eastern Mare Imbrium, notable for its almost perfectly flat floor.

Aristillus Prominent crater 55 km in diameter in eastern Mare Imbrium with terraced walls, numerous surrounding ridges and a complex central peak 900 m high. Under high illumination it is seen surrounded by a faint ray system. Forms a pair with Autolycus to the south.

Aristoteles Magnificent, partially lava-flooded crater 87 km in diameter, touching the smaller crater Mitchell to the east. Numerous ridges radiate from its outer walls. Makes a pair with Eudoxus to the south.

Autolycus Prominent crater 39 km in diameter, south of Aristillus. High illumination shows it to be the centre of a faint ray system.

Cassini Flooded ring of unusual appearance with partially destroyed walls, 56 km in diameter. Contains a bowl-shaped crater, Cassini A, 17 km in diameter.

Eratosthenes Prominent, deep crater on the edge of Sinus Aestuum at the southern end of the Apennine Mountains. Terraced walls and craterlet on central peak. Diameter 58 km.

Eudoxus Rugged crater 67 km in diameter with small central peak and terraced walls. South of Aristoteles.

Godin Smaller but deeper companion to Agrippa; 35 km in diameter with central peak, bright walls and faint ray system.

Hyginus Rimless crater 10 km in diameter, centre of a cleft or rille, 225 km long, visible in small telescopes, evidently formed by collapse of the surface. To the east is another cleft, the Ariadaeus Rille.

Lambert Crater 30 km in diameter in Mare Imbrium, with central craterlet. Situated on a wrinkle ridge. Under low illumination a larger 'ghost' ring, Lambert R, is seen to the south.

Linné Bright spot on Mare Serenitatis, best seen under high illumination. At its centre is a small, young crater 2.4 km in diameter.

Manilius Bright crater 39 km in diameter in Mare Vaporum, with terraced walls and central peak. Develops a ray system as illumination increases, as does its neighbour Menelaus (diameter 27 km) on the rim of Mare Serenitatis.

Mare Imbrium Enormous circular plain 1250 km in diameter, dominating this section. Bounded by the Alps, Caucasus, Apennine and Carpathian Mountains but open on the southwest to Oceanus Procellarum. Mare Imbrium has a double structure: traces of a smaller inner ring are visible, marked by a few isolated mountains and wrinkle ridges. Note different shades of lava on its surface. Isolated mountains protruding from its dark floor include Pico, Piton, the Straight Range, the Teneriffe Mountains and the Spitzbergen Mountains.

Mare Serenitatis A major, rounded lunar sea 680 × 580 km, bounded on the northwest by the Caucasus Mountains and on the southwest by the Haemus Mountains. A bright ray from Tycho crosses the dark lava plain, passing through the crater Bessel, 16 km in diameter. Under high illumination, Mare Serenitatis is seen to be rimmed with darker lava. Note also a major wrinkle ridge (the Serpentine Ridge) on the east side. Apollo 17 landed at the southeastern edge of Mare Serenitatis.

Plato Unmistakable large, dark floored crater in the highlands north of Mare Imbrium, 100 km in diameter. Prominent under all conditions of illumination. Tiny craterlets pockmark its flat floor. Temporary obscurations of the surface, presumed to be caused by outgassing, have been observed in this area. Landslips appear to have detached part of the inner western wall.

Pytheas Small (20 km diameter) but deep and prominent crater on Mare Imbrium, rhomboidal in outline. Becomes brilliant under high illumination.

Stadius 'Ghost' ring to the east of Copernicus, outlined only by a few ridges and craterlets. Visible only under low illumination. Diameter 64 km.

Timocharis Bright crater 35 km in diameter on Mare Imbrium, with terraced walls and distinctive central crater. Faint ray centre.

Triesnecker Crater 26 km in diameter surrounded by a system of clefts.

Moon Map 3

Atlas Large crater, 87 km in diameter, with terraced walls and complex floor. A ruined ring abuts to the northwest. Atlas and Hercules form one of the many crater pairs in this region.

Burckhardt Complex elliptical (45 × 55 km) crater, appearing to overlap older formations on either side.

Bürg Prominent crater despite its moderate size (diameter 40 km). Central peak. Lies in the centre of Lacus Mortis. Note nearby rilles.

Cleomedes Large, irregular-shaped crater 126 km in diameter with partially flooded floor, to north of Mare Crisium. West wall is interrupted by 43 km diameter crater Tralles.

Endymion Large, dark-floored enclosure 125 km in diameter, with walls up to 4900 m high.

Franklin Crater 56 km in diameter. Forms a pair with the 40 km diameter Cepheus to the northwest.

Geminus Prominent crater 86 km in diameter with central peak. Bright rays emanate from two smaller craters nearby, Messala B and Geminus C.

Hercules Flat floored enclosure 67 km in diameter containing sharp, bright bowl crater Hercules G.

Lemonnier Old and flooded crater with its western wall washed away by lava from Mare Serenitatis. Diameter 61 km.

Mare Crisium Unmistakable dark lowland plain ringed by high mountains like an oversized crater, 435 × 565 km. Main feature on its flat, lava-flooded floor is the crater Picard, 23 km in diameter, with smaller crater Peirce to its north.

Mare Tranquillitatis An irregular-shaped lowland 650 × 900 km. Wrinkle ridges attest to numerous lava flows over the plain. Main crater on the Mare is the distorted Arago, on the western side, 26 km in diameter. Apollo 11 made the first manned lunar landing in the southwest corner of Mare Tranquillitatis.

Plinius Distinctive crater 43 km in diameter with complex central peak combined with a central craterlet. Stands between Mare Serenitatis and Mare Tranquillitatis, along with Dawes to the northeast, 18 km in diameter.

Posidonius Large (diameter 100 km) partially flooded and ruined crater on the northeast shore of Mare Serenitatis. Its floor contains a curving ridge, several rilles and a small bowl crater. Ruined structure Chacornac, 51 km in diameter, abuts to the southeast.

Proclus Small (diameter 28 km) but brilliant crater on the western edge of Mare Crisium. High illumination shows it to be the centre of a fan-shaped ray system.

Taruntius Low-walled crater in northwestern Mare Fecunditatis, 56 km in diameter, with concentric inner ring. Crater Cameron (formerly Taruntius C) interrupts the northwestern wall. Centre of a faint ray system.

Thales Bright ray crater 32 km in diameter, northeast of Mare Frigoris.

Moon Map 4

Bullialdus Handsome crater in Mare Nubium with terraced walls and complex central peak. Diameter 59 km. Two smaller craters, Bullialdus A and B, form a chain extending south.

Flamsteed Small (diameter 21 km) crater on Oceanus Procellarum with much larger ring of eroded hills to its north, Flamsteed Px.

Gassendi Large, 100 km in diameter, partially flooded ring on the northern border of Mare Humorum. Complex internal pattern of clefts, ridges and hillocks. A rich area for transient lunar events (TLPs). Deeper crater Gassendi A interrupts the north wall, with the smaller Gassendi B lying further north.

Grimaldi Vast 222 km diameter dark floored enclosure at the western limb of the Moon with broad, crater-strewn walls. Nearer the limb is a smaller dark patch, marking the floor of Riccioli, 152 km in diameter.

Hainzel Curious keyhole-shaped crater, 92 × 66 km, evidently composed of three craters fused together.

Hippalus Lava-flooded bay 58 km in diameter on shores of Mare Humorum, in an area with many rilles.

Lansberg Prominent crater 40 km in diameter with massive walls and a central peak, on Oceanus Procellarum.

Letronne Large bay, 119 km wide, on the south shore of Oceanus Procellarum. Its seaward facing wall has evidently been washed away by invading dark lava.

Mare Humorum Rounded lowland plain 420 km across with irregular borders. Ringed by clefts and wrinkle ridges. On the south it invades the rings Doppelmayer and Lee, although Vitello escapes destruction. On the east is the ringed bay Hippalus, associated with much surface faulting.

Schickard Major dark floored enclosure 227 km in diameter. South of Schickard are the overlapping craters Nasmyth (diameter 77 km) and Phocylides (diameter 114 km). Adjoining it on the southwest is the extraordinary plateau Wargentin, 84 km across, evidently a crater filled to the brim with solidified lava.

Schiller Curious footprint-shaped enclosure, 179×71 km.

Sirsalis and Sirsalis A Twin craters near to a long cleft, Rima Sirsalis, which stretches towards Darwin.

Moon Map 5

Abulfeda Prominent, smooth floored crater with sculptured inner walls, 62 km in diameter. Apollo 16 landed in the highlands north of here in 1972.

Albategnius Large (diameter 136 km) walled enclosure with central peak. The prominent crater Klein, 44 km in diameter, breaks the southwest wall.

Aliacensis Prominent crater with irregular outline, 80 km in diameter. Forms a pair with Werner.

Alpetragius 3900 m deep bowl-shaped crater on outer slopes of Alphonsus with large central dome. Diameter 40 km.

Alphonsus Large enclosure 118 km in diameter with complex walls and a central ridge. Numerous craterlets and clefts cover the floor; several dark patches are visible under high illumination. Alphonsus is the site of reported obscurations believed to be due to release of gas from the surface.

Arzachel Magnificent crater 97 km in diameter with terraced walls and prominent central peak. To its east is a noticeable crater 32 km in diameter with central peak, like a smaller version of Alpetragius, called Parrot C.

Barocius Large formation southeast of Maurolycus, diameter 82 km. Its northeast wall is broken by Barocius B. To the southwest is Clairaut, 75 km in diameter, between Barocius and Cuvier.

Birt Sharp, bright crater 17 km in diameter on eastern Mare Nubium. Telescopes reveal a smaller crater (diameter 7 km), Birt A, on the east wall and, under low illumination, a rille to the west.

Blancanus Crater 105 km in diameter, south of Clavius.

Clavius Magnificent walled plain 225 km in diameter. Note the distinctive arc of smaller craters across its convex floor. Its south wall is interrupted by the 50 km crater Rutherfurd, and its northeast wall by 52 km Porter.

Delambre Prominent crater 53 km in diameter with irregular interior, southwest of Mare Tranquillitatis.

Deslandres Huge, low and eroded formation 234 km in diameter, southeast of Mare Nubium. The partly ruined ring Lexell, diameter 63 km, opens onto its southern side, and the prominent crater Hell (diameter 33 km) lies on its western floor.

Fra Mauro Largest member, 94 km wide, of old and eroded crater group north of Mare Nubium, also including Bonpland (diameter 60 km), Parry (diameter 46 km), and Guericke (diameter 58 km). Apollo 14 landed at Fra Mauro in 1971.

Heraclitus Curious elongated formation with central ridge, south of Stöfler. Its southern end is rounded off by the crater Heraclitus D. Between Heraclitus and Stöfler is the crater Licetus, diameter 75 km. Touching Heraclitus to the east is Cuvier, also 75 km in diameter.

Herschel Deep (3900 m) crater with elongated central peak, north of Ptolemaeus; diameter 41 km. Farther north is the slightly smaller Spörer, a partially filled ring.

Hipparchus Large, eroded enclosure 150 km in diameter north of Albategnius. Central 'peak' is actually a small ruined crater. On its northeastern floor is Horrocks, 30 km in diameter and 2800 m deep. Between Hipparchus and Albategnius lies the crater Halley (diameter 36 km), east of which is Hind, 29 km wide and 2800 m deep.

Longomontanus Large walled plain in the rugged southern uplands of the Moon. Diameter 145 km. A ridge to the east forms a crescent-shaped enclosure known as Longomontanus Z.

Maginus Major walled plain 163 km in diameter north of Clavius. Convex floor with small peaks in centre. Southwest wall is interrupted by smaller crater, Maginus C.

Mare Nubium Irregular dark lowland plain, covered with numerous wrinkle ridges and ghost craters. On its southern shore is the flooded crater Pitatus, and on the southwest are the dark floored pair Campanus and Mercator (48 km and 47 km in diameter). Its most celebrated feature is a 26 km long fault on the eastern side called the Rupes Recta, or Straight Wall, between the craters Birt and Thebit. The Straight Wall appears to run north–south through the remains of an old flooded crater of which only the eastern half remains. At the southern end of the Straight Wall are the Stag's Horn Mountains, apparently the remains of a flooded crater.

Maurolycus Distinctive crater, diameter 114 km, with twin central peak. Its walls rise to 5000 m. It partially obliterates a smaller, unnamed formation to the north.

Moretus Crater with a central peak in the jumbled uplands southeast of Clavius. Diameter 114 km. Closer to the south pole are the craters Short (diameter 71 km) and Newton (diameter 64 km), heavily foreshortened.

Pitatus Large, dark floored ring 105 km across on the southern shore of Mare Nubium. Invading lava from Mare Nubium has partly destroyed the crater's walls and left only the vestige of a central peak. Note rilles around its inner walls. A smaller, similarly flooded ring adjoining it on the east is Hesiodus.

Ptolemaeus Vast walled plain 153 km in diameter, hexagonal in shape. Its ancient floor is heavily pockmarked with smaller craters, the most prominent being Ptolemaeus A.

Purbach Battered but still prominent large crater, 118 km in diameter. Its floor contains ridges, and its north wall is interrupted by the oval crater Purbach G, while its south wall intrudes into Regiomontanus.

Regiomontanus Eroded crater 126 × 110 km. Central peak has summit craterlet. Forms a pair with Purbach, both noticeably hexagonal in outline.

Scheiner Crater 110 km in diameter southwest of Clavius. The largest of three craterlets on its floor is Scheiner A.

Stöfler Large, flat floored formation 137 km wide, west of Maurolycus. Its eastern wall is destroyed by the intrusion of several craters, the largest being Faraday, diameter 69 km. The southern wall of Faraday is disturbed by Faraday C, which itself intrudes into Stöfler P.

Thebit Fascinating triple crater on southeastern Mare Nubium. The main crater, 55 km in diameter, is broken by the 20 km Thebit A, which in turn is broken by the smaller Thebit L.

Tycho Magnificent crater in the Moon's southern uplands, 85 km in diameter. Prominent at all angles of illumination, and brilliant under high lighting. Massively terraced walls rising up to 4500 m, imposing central peak and rough floor. Tycho is the major ray crater on the Moon. Rays from Tycho extend for 1500 km or more in all directions. Note dark 'collar' around Tycho under high Sun. Probably the youngest of the Moon's major features.

Walter Large crater 135 km in diameter, considerably modified by landslips on the inner walls and by several interior craterlets. Appears almost square in outline.

Werner Prominent crater 70 km in diameter with walls 4200 m high, notably sharper and more rounded than its neighbouring craters. The floor of Werner is dotted with several hills.

Moon Map 6

Capella Prominent crater north of Mare Nectaris, 45 km in diameter, deformed by a surface fault. Large central peak. Isidorus, 39 km in diameter, adjoins it to the west.

Catharina Member of a curving trio of craters around western Mare Nectaris. Diameter 97 km. A faint ring, Catharina P, covers a large part of its northern floor.

Cyrillus Crater 93 km in diameter with complex terraced walls, multiple central peak and rugged floor. Overlapped by Theophilus.

Fracastorius Horseshoe-shaped bay 124 km in diameter on southern shore of Mare Nectaris. Dark lava has breached its northern wall and flooded its interior.

Janssen Vast, irregular enclosure, 190 km in diameter, heavily bombarded. In the north it is interrupted by Fabricius, 78 km in diameter, which has a central peak. On the west wall is a smaller (diameter 34 km) sharp crater, Lockyer. Southeast of Janssen are the twin craters Steinheil and Watt (67 km and 66 km in diameter respectively). Farther north of Fabricius is Metius, diameter 88 km.

Langrenus Magnificent, bright, walled plain on eastern Mare Fecunditatis with terraced walls, outer ridges and a complex central peak. Its diameter is 132 km. Langrenus is a ray centre. Northwest of it, on Mare Fecunditatis, is a trio of smaller craters called Langrenus F, B and K in order of decreasing size. South of Langrenus is the large, flooded formation Vendelinus, 147 km in diameter.

Mädler Prominent crater (diameter 28 km) on northwest Mare Nectaris, with a central ridge.

Mare Fecunditatis Irregularly-shaped dark lowland area, connecting with Mare Tranquillitatis. On its western border it invades several craters, notably Gutenberg (diameter 71 km) and Goclenius (diameter 60 km). Note numerous clefts in this area.

Mare Nectaris Rounded lowland plain 400 km wide, bordered by several large craters, notably Theophilus, Cyrillus, Catharina and Fracastorius. An outer mountain ring, the Altai scarp, surrounds Mare Nectaris.

Messier and Messier A Elliptical pair of craters on Mare Fecunditatis, prominent despite their small sizes of 9×11 km and 13×11 km. From the western member, Messier A, extend two bright rays. Both craters appear brilliant under high illumination.

Palitzsch A crater and valley on the eastern side of Petavius. The crater itself is at the southern end of the valley, which is 150 km long.

Petavius Magnificent walled enclosure 177 km in diameter. A prominent rille runs across the floor from the massive, complex central peak to the terraced walls which appear double in parts. Ridges radiate from its outer walls. West of Petavius is Wrottesley, 57 km in diameter and with a central peak. South of it is the 83 km diameter Hase.

Piccolomini Beautiful crater on the Altai scarp. Diameter 89 km, with a broad central peak and terraced walls.

Rheita Valley Crater chain northeast of Janssen and Fabricius. It can be traced for a total length of about 450 km. The crater Rheita itself, 70 km in diameter and with a small central peak, lies on the northern end of the valley.

Theophilus Imposing crater 100 km in diameter on the northwestern rim of Mare Nectaris, with a massive 2200 m central mountain. Terraced walls rise over 5000 m above the floor, with many external ridges.

315

Map 1

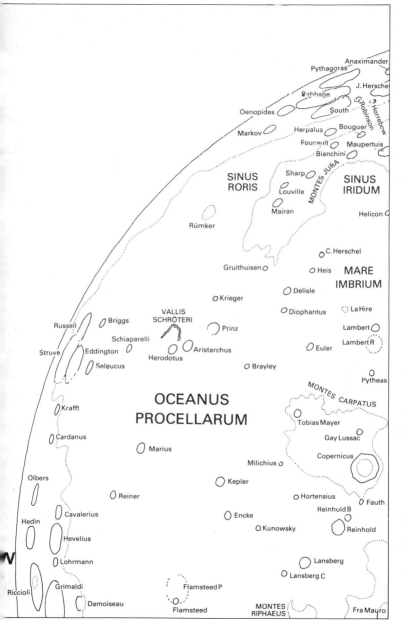

Anaximander
Pythagoras
J. Herschel
Babbage
Robinson
Oenopides
South
Horrebow
Markov
Harpalus
Bouguer
Foucault
Maupertuis
Bianchini
SINUS
RORIS
Sharp
MONTES JURA
SINUS
IRIDUM
Louville
Mairan
Helicon
Rümker
C. Herschel
Gruithuisen
Heis
MARE
IMBRIUM
Delisle
Krieger
VALLIS
SCHRÖTERI
Diophantus
La Hire
Russell
Briggs
Prinz
Lambert
Schiaparelli
Aristarchus
Lambert R
Struve
Eddington
Herodotus
Seleucus
Euler
Brayley
Pytheas
MONTES CARPATUS
Krafft
OCEANUS
PROCELLARUM
Tobias Mayer
Cardanus
Gay Lussac
Marius
Copernicus
Milichius
Olbers
Kepler
Reiner
Hortensius
Fauth
Cavalerius
Encke
Reinhold B
Hedin
Kunowsky
Reinhold
Hevelius
Lansberg
Lohrmann
Lansberg C
Riccioli
Grimaldi
Flamsteed P
Damoiseau
Flamsteed
MONTES
RIPHAEUS
Fra Mauro

W

Map 2

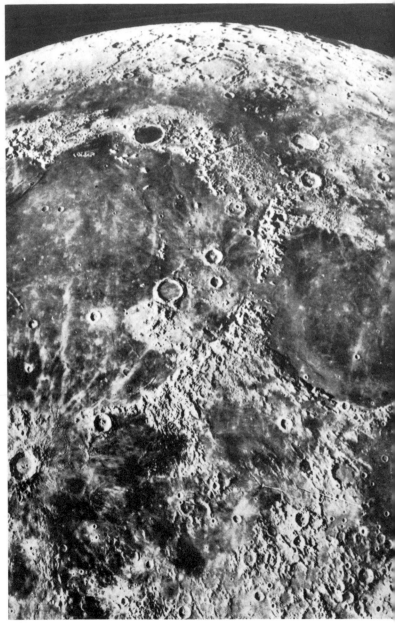

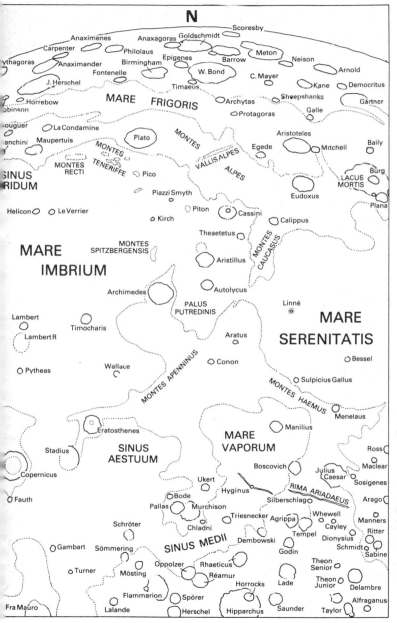

Map 3

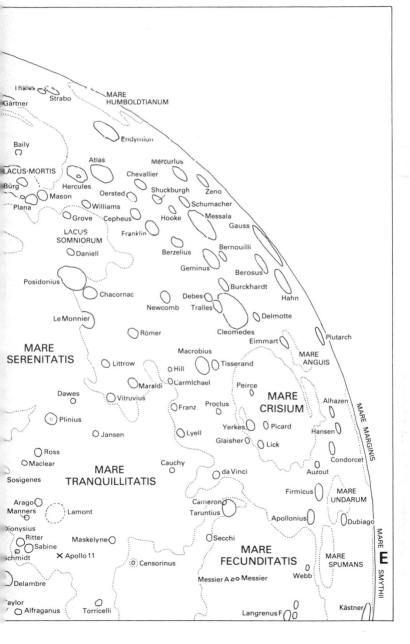

Thales
Gärtner
Strabo
MARE HUMBOLDTIANUM
Endymion
Baily
Atlas
Mercurlus
LACUS·MORTIS
Chevallier
Bürg
Hercules
Shuckburgh
Zeno
Mason
Oersted
Plana
Schumacher
Williams
Grove
Cepheus
Hooke
Messala
Franklin
Gauss
LACUS SOMNIORUM
Daniell
Berzelius
Bernouilli
Geminus
Posidonius
Berosus
Chacornac
Burckhardt
Debes
Hahn
Newcomb
Tralles
Le Monnier
Delmotte
Römer
Cleomedes
Eimmart
Plutarch
MARE SERENITATIS
Macrobius
MARE ANGUIS
Littrow
Hill
Tisserand
Maraldi
Carmichael
Peirce
Dawes
Vitruvius
Franz
Proclus
MARE CRISIUM
Alhazen
Plinius
Jansen
Lyell
Yerkes
Picard
Hansen
Glaisher
Lick
Ross
Maclear
Cauchy
Condorcet
Sosigenes
MARE TRANQUILLITATIS
da Vinci
Auzout
Arago
Firmicus
MARE UNDARUM
Manners
Lamont
Cameron
Dionysius
Taruntius
Apollonius
Dubiago
Ritter
Maskelyne
Sabine
Schmidt
Apollo 11
Secchi
MARE FECUNDITATIS
MARE SPUMANS
Delambre
Censorinus
Webb
MARE SMYTHII
Taylor
Alfraganus
Torricelli
Messier A
Messier
Kästner
Langrenus F
MARE MARGINIS
E

Map 4

N

OCEANUS PROCELLARUM

MARE ORIENTALE

MARE COGNITUM

MONTES RIPHAEUS

MARE HUMORUM

PALUS EPIDEMIARUM

Olbers, Cavalerius, Reiner, Encke, Fauth, Reinhold B, Reinhold, Kunowsky, Hevelius, Hedin, Lohrmann, Lansberg, Lansberg C, Riccioli, Flamsteed P, Flamsteed, Grimaldi, Damoiseau, Euclides, Fra Mauro, Letronne, Bonpland, Hansteen, Sirsalis, Rocca, Sirsalis A, Billy, Herigonius, Gassendi B, Gassendi A, Darney, Fontana, Gassendi, Crüger, Lubiniezky, Darwin, de Vico, Mersenius, Agatharchides, Bullialdus, Prosper Henry, Bullialdus A, Bullialdus B, Cavendish, Loewy, König, Eichstädt, Liebig, MARE HUMORUM, Hippalus, Campanus, Kies, Lamarck, Paul Henry, de Gasparis, Doppelmayer, Mercator, Byrgius, Vieta, Palmieri, Puiseux, Krasnov, Fourier, Lee, Vitello, Ramsden, Cichus, Elger, Capuanus, Clausius, Haidinger, Lehmann, Drebbel, Hainzel, Epimenides, Inghirami, Schickard, Mee, Wargentin, Nasmyth, Nöggerath, Bayer, Phocylides, Schiller, Rost, Weigel, Zucchius, Bettinus

Map 5

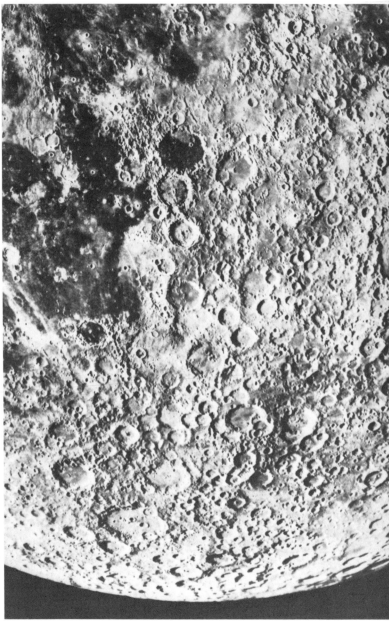

South Central Section

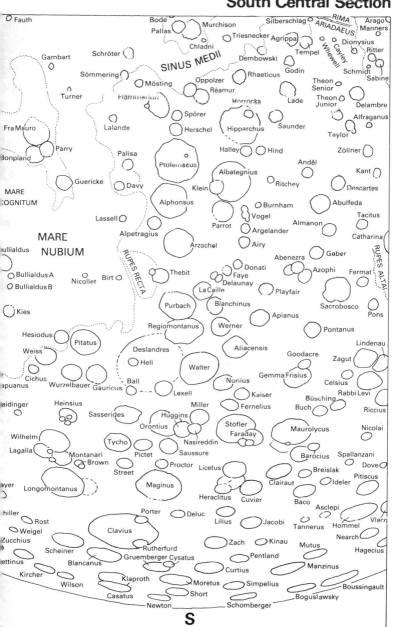

Fauth

Bode
Pallas
Murchison
Chladni
Silberschlag
RIMA ARIADAEUS
Arago
Manners
Whewell
Cayley
Dionysius
Ritter
Triesnecker
Agrippa
Tempel
Schröter
Dembowski
Godin
Schmidt
Sabine
Gambart
SINUS MEDII
Sömmering
Mösting
Oppolzer
Rhaeticus
Theon Senior
Theon Junior
Delambre
Turner
Flammarion
Réamur
Lade
Horrocks
Spörer
Herschel
Hipparchus
Saunder
Alfraganus
Taylor
Fra Mauro
Lalande
Halley
Hind
Zöllner
Parry
Palisa
Ptolemaeus
Andèl
Kant
Bonpland
Guericke
Davy
Klein
Albategnius
Ritchey
Descartes
MARE COGNITUM
Alphonsus
Burnham
Abulfeda
Lassell
Vogel
Almanon
Tacitus
Alpetragius
Parrot
Argelander
Catharina
Arzachel
Airy
Geber
MARE NUBIUM
Abenezra
Azophi
Fermat
Bullialdus
RUPES RECTA
Thebit
Donati
Faye
Delaunay
Playfair
RUPES ALTAI
Bullialdus A
Nicollet
Birt
La Caille
Bullialdus B
Purbach
Blanchinus
Sacrobosco
Pons
Kies
Regiomontanus
Werner
Apianus
Pontanus
Hesiodus
Pitatus
Deslandres
Aliacensis
Lindenau
Weiss
Hell
Walter
Goodacre
Zagut
Cichus
Ball
Nonius
Gemma Frisius
Celsius
Wurzelbauer
Gauricus
Lexell
Kaiser
Rabbi Levi
Capuanus
Büsching
Heinsius
Miller
Fernelius
Buch
Riccius
Haidinger
Sasserides
Huggins
Stöfler
Maurolycus
Nicolai
Wilhelm
Orontius
Faraday
Lagalla
Tycho
Nasireddin
Barocius
Spallanzani
Montanari
Pictet
Saussure
Breislak
Dove
Brown
Proctor
Licetus
Clairaut
Ideler
Pitiscus
Street
Baco
Longomontanus
Maginus
Heraclitus
Cuvier
Asclepi
Schiller
Porter
Deluc
Lilius
Jacobi
Tannerus
Hommel
Vlacq
Rost
Weigel
Clavius
Zach
Kinau
Mutus
Nearch
Zucchius
Hagecius
Bettinus
Scheiner
Rutherfurd
Gruemberger
Cysatus
Pentland
Manzinus
Kircher
Blancanus
Curtius
Wilson
Klaproth
Moretus
Simpelius
Boussingault
Casatus
Short
Boguslawsky
Newton
Schomberger

S

Map 6

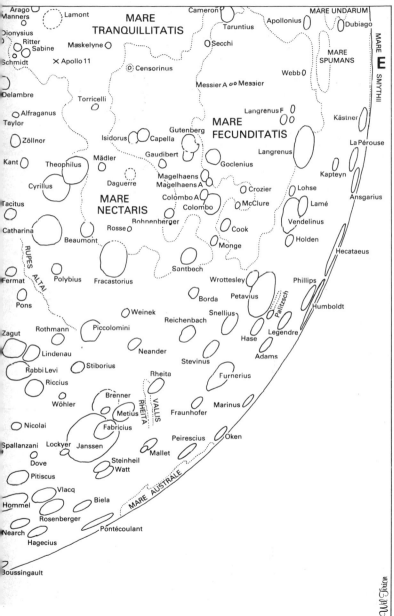

MARE TRANQUILLITATIS

Arago
Manners
Lamont
Cameron
MARE UNDARUM
Taruntius
Apollonius
Dubiago
Dionysius
Ritter
Sabine
Maskelyne
Secchi
MARE SPUMANS
MARE E SMYTHII
Schmidt
Apollo 11
Censorinus
Webb
Delambre
Messier A
Messier
Alfraganus
Taylor
Torricelli
Langrenus F
Kästner
Zöllner
Isidorus
Capella
Gutenberg
MARE FECUNDITATIS
La Pérouse
Kant
Theophilus
Mädler
Gaudibert
Langrenus
Cyrillus
Daguerre
Magelhaens
Magelhaens A
Goclenius
Kapteyn
Colombo A
Crozier
Lohse
Ansgarius
MARE NECTARIS
Colombo
McClure
Lamé
Tacitus
Rohnenberger
Vendelinus
Catharina
Rosse
Cook
Beaumont
Monge
Holden
Hecataeus
Fermat
Polybius
Fracastorius
Santbech
Wrottesley
Phillips
RUPES ALTAI
Borda
Petavius
Pons
Weinek
Snellius
Palitzsch
Humboldt
Zagut
Rothmann
Piccolomini
Reichenbach
Hase
Legendre
Lindenau
Neander
Stevinus
Adams
Rabbi Levi
Stiborius
Riccius
Rheita
Furnerius
Wöhler
Brenner
VALLIS RHEITA
Nicolai
Metius
Fraunhofer
Marinus
Fabricius
Spallanzani
Lockyer
Janssen
Peirescius
Oken
Dove
Mallet
Pitiscus
Steinheil
Watt
Vlacq
Biela
MARE AUSTRALE
Hommel
Nearch
Rosenberger
Pontécoulant
Hagecius
Boussingault

Eclipses of the Sun and Moon

Occasionally the Sun, Moon and Earth line up exactly to cause an eclipse. When the Moon comes between the Sun and the Earth it blocks off the Sun's light from part of the Earth, causing an eclipse of the Sun. When the Moon is on the opposite side of the Earth to the Sun, the Earth can block off the Sun's light from the Moon, causing an eclipse of the Moon. If the Moon orbited in the same plane as the Earth's orbit around the Sun, then an eclipse would occur at each full or new Moon. But since the Moon's orbit is inclined 5° with respect to the orbit of the Earth, eclipses occur only on those occasions when the Moon is crossing the Earth's orbit at new or full Moon.

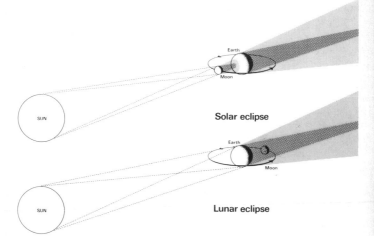

When the Moon passes in front of the Sun as seen from Earth, it causes a solar eclipse. When the Moon enters the Earth's shadow, a lunar eclipse occurs. *Wil Tirion.*

The maximum number of eclipses possible in one year is seven. Each year there are at least two eclipses of the Sun and two of the Moon, although they will not all be visible from one place on Earth. An eclipse of the Moon is visible wherever the Moon is above the horizon, but an eclipse of the Sun can be seen only from within the narrow band on which the Moon's shadow falls.

Scientifically, total eclipses of the Sun are by far the most important. At a total solar eclipse the Moon completely blots out the brilliant disk of the Sun, allowing astronomers to observe the Sun's faint outer halo of gas known as the corona. (It is a fortunate coincidence that the Sun and Moon appear almost exactly the same size in the sky.) Astronomers travel across the globe for the few precious moments that totality affords. The longest that a total eclipse of the Sun can last is 7 minutes 40 seconds, but the usual duration is 2–3 minutes.

The band of totality is a mere 300 km across at its widest, but outside this band is a much wider area in which a partial eclipse of the Sun can be seen. Sometimes, when the Moon is at its farthest from us in its elliptical orbit, it is too small to cover the Sun's disk completely, and so a ring of bright sunlight remains visible around the dark outline of the

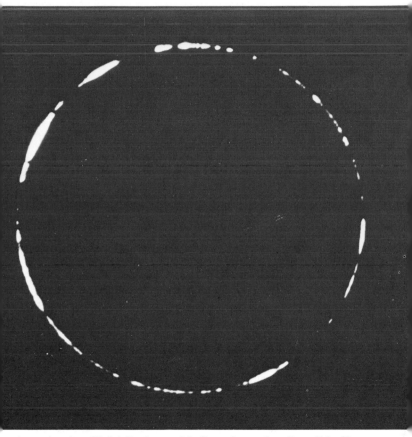

A complete ring of Baily's Beads around the Sun at the annular eclipse of May 20, 1966, as seen from the Greek island of Lesvos. This diamond necklace effect was caused by mountains around the rim of the Moon. *John Mathers*.

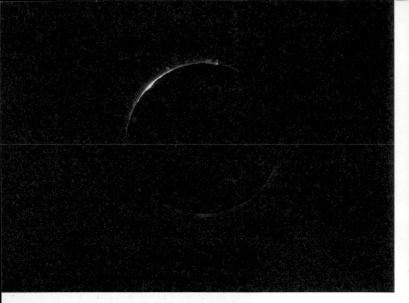

Above: Prominences of hydrogen gas glow pinkly at the edge of the Sun, as seen from Zanzibar during the total solar eclipse of October 23, 1976. *Paul Cass.*

Left: A crescent Sun: over half the Sun's disk is blotted out by the Moon at the partial solar eclipse of February 25, 1971, seen through clouds. *Ian Ridpath.*

Moon. Such an eclipse is termed an *annular* eclipse (from the Latin word *annulus* meaning a ring, not because they occur yearly).

Partial and annular eclipses are of curiosity value only; they have none of the scientific importance of a total eclipse. A solar eclipse starts at *first contact*, when the edge of the Moon begins its progress across the face of the Sun. Totality is still 1½ hours away. The partial phases of the eclipse can be observed by looking at the Sun through a dark filter, or by projecting the Sun's image with binoculars or telescope onto a white card. When you do this, compare the jet black outline of the Moon with the umbra of sunspots. This will confirm that sunspots are not totally black; instead their colour appears somewhat brownish.

A suitable filter for looking at the Sun is heavily overexposed black and white photographic negative film (use two or three layers if necessary). You can look at the Sun quite safely through this, because the silver in the film absorbs both the light and the heat from the Sun. Colour film is not suitable as a filter because the colour dyes do not absorb infrared (heat) radiation; although colour film dims the light from the Sun sufficiently, the Sun's heat can still get through to damage your eyes. Another simple but safe way of observing an eclipse of the Sun is to make a pinhole in a piece of card and allow the Sun's light to pass through this hole onto a white surface. You are in fact using a pinhole camera to observe the Sun. The drawback with this system is that the image so formed is small and faint.

Not until about 15 minutes before totality, when the Sun's disk is more than 80 per cent covered, does the sky start to get noticeably dark. An eerie half-light falls over the landscape; animals act as though night is falling. Totality itself rushes up as the last crescent sliver of sunlight is blotted out by the Moon. In the final seconds, chinks of sunlight peep between the mountains at the Moon's rugged edge, causing the so-called Baily's Beads, named after the English astronomer Francis Baily who described them after the eclipse of 1836. Often one bead shines brighter than the others, producing the effect of a diamond ring.

Then, second contact: the Moon completely covers the Sun, and the pearly coloured corona springs into view. Featherlike plumes and streamers of the corona extend outwards from the polar and equatorial regions of the Sun for several solar diameters. Pinkish-red prominences can be seen looping out from the Sun's chromosphere around the dark outline of the Moon. Bright stars are visible in the darkened sky. All too soon, the beautiful spectacle is over. The diamond ring flashes out at third contact, signalling the end of totality. The Moon moves clear of the Sun. The eclipse is over.

Eclipses of the Moon are far less spectacular. The Moon takes several hours to move completely through the dark inner part of the Earth's shadow, the umbra. The outer part of the shadow, the penumbra, is so light that it produces no noticeable darkening of the Moon's surface.

A total eclipse of the Moon can last up to $1\frac{3}{4}$ hours, but even when totally eclipsed the Moon seldom completely disappears. The reason for this is that light is bent into the Earth's shadow by the Earth's atmosphere, giving the eclipsed Moon a coppery-red colour. Dark eclipses occur when there are a lot of clouds and dust in the Earth's atmosphere which serve to block the light. Although of interest as a natural spectacle, there is little of scientific importance in a lunar eclipse.

Mercury

Mercury is a disappointing object for observers. Through even the largest telescopes it appears as nothing more than an almost featureless orange blob, less distinct than the Moon appears to the naked eye, that goes through phases as it orbits the Sun every 88 days. Most astronomers must therefore be content with simply catching a glimpse of this elusive object during one of its periodic appearances in the morning or evening sky.

Because Mercury is the closest planet to the Sun, it and the Sun never appear far apart in the sky. Circumstances dictate that there are two good times to look for Mercury. One is when it is setting after the Sun during evenings in spring (around March–April in the northern hemisphere or September–October in the southern hemisphere). The second is when it is a morning object rising before the Sun in the autumn (September–October in the northern hemisphere, March–April in the southern hemisphere).

An additional complication is that its orbit is markedly elliptical, ranging from 46–70 million km from the Sun, so that even on these occasions there are times when Mercury is easier to see than on others. Even when Mercury is best placed, a low, clear horizon will be needed to see it; binoculars help to pick the planet out of the twilight glow, for Mercury can never be seen against a truly dark sky. With all these complications, it is little wonder that many town dwellers have never seen the planet. Nevertheless it is worth looking for, because at its brightest it can shine at magnitude −1 or more.

The difficulty in observing Mercury led to a long-standing mistake concerning the time it takes to rotate on its axis. Towards the end of the 19th century, the Italian astronomer Giovanni Schiaparelli proposed, after a long series of observations, that the planet spins on its axis in 88 days, the same time as it takes to orbit the Sun. It would therefore keep one face turned permanently towards the Sun, as the Moon does to the Earth. In the 1920s the Greek-born astronomer Eugene Antoniadi compiled a map showing smudgy markings on the surface of Mercury based on the assumption of an 88 day rotation period. This map seemed to settle the matter once and for all.

Then, in 1965, came a surprise. At Arecibo Radio Observatory, astronomers Rolf Dyce and Gordon Pettengill bounced radio waves off the surface of Mercury. From the change in frequency of the reflected radio waves, they deduced that Mercury spins once every 59 days, two-thirds of the time that it takes to orbit the Sun. Therefore the Sun does rise and set on Mercury, but very slowly. For the Sun to go once around the sky as seen from the surface of the planet – say, from

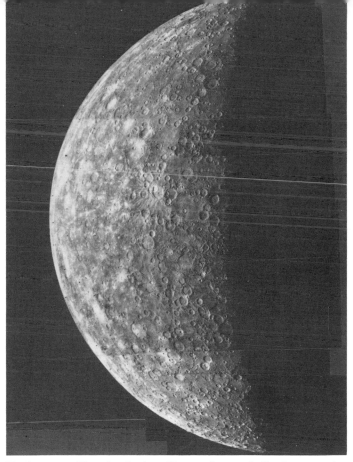

Not the crescent Moon, but the planet Mercury photographed by the space probe Mariner 10 in March 1974. The bright ray crater just above centre is called Kuiper. The largest craters are about 200 km across. *Jet Propulsion Laboratory*.

one noon to the next – takes 176 Earth days, during which time Mercury orbits the Sun twice, spinning three times on its axis.

In the skies of Mercury the Sun appears 2.5 times as large as it does from Earth. The Sun's intense heat roasts the surface rocks of Mercury to over 400°C at noon on the equator, hot enough to melt tin and lead. Without an atmosphere to hold in the heat, the planet's surface cools to a frigid −170°C during the long night. The daytime side of Mercury is continually blasted by lethal doses of high–energy solar radiation.

Astronomers had long assumed that Mercury resembled our Moon in appearance. For one thing, it is only 50 per cent larger than our Moon, with a diameter of 4880 km. The only planet in the Solar System smaller than that is Pluto. But it took the space probe Mariner 10 in 1974 to show how remarkably similar Mercury and the Moon appear. As Mariner 10 flew past Mercury, its cameras photographed a surface heavily pockmarked with craters of all sizes, similar to the lunar highlands. Mariner 10 photographed little more than a third of Mercury's surface yet, assuming that portion to be typical of the rest of the planet, it is enough to tell us much of the previously unknown story of Mercury.

Craters on Mercury look almost identical to their lunar counterparts. There are deep, young craters, eroded ancient craters, craters with terraced walls, central peaks and bright rays. Many of the features on Mercury have been named after artists, composers and writers, thereby breaking the near-monopoly that astronomers' names have held on the surfaces of Solar System bodies. For instance, we now find on Mercury memorials to Bach, Mozart, Van Gogh and Chekhov.

Sometimes it is difficult to distinguish at a glance between a picture of Mercury and one of the Moon. Almost certainly, the craters on both bodies have been formed in the same way, by the impact of large meteorites early in the history of the Solar System. One noticeable difference on Mercury is that material ejected from the craters has not travelled as far as that on the Moon, because Mercury's surface gravity is stronger – over twice that of the Moon, but still only 38 per cent that of the Earth. Another result of the higher gravity of Mercury is that craters tend to be shallower for a given diameter than on the Moon.

Mercury has cliffs hundreds of kilometres long and 2–4 km high, unlike anything on the Moon. These are believed to have been caused by a shrinking of the planet as its core cooled early in its history, leading to compression and faulting in the crustal rocks. The surface rocks of Mercury are actually slightly darker in colour than those of the Moon, reflecting a mere 6 per cent of the sunlight hitting them, compared with 7 per cent for the Moon; Mercury in fact has the darkest surface of any planet in the Solar System.

Between many of the large craters on Mercury are areas of ancient crust peppered only with small craters. These areas, termed the inter-crater plains, have no real counterpart on the Moon. They clearly pre-date the large craters, and may be examples of the original crust of Mercury. By contrast, the Moon's original crust has been totally churned up by impacts so that none of it remains unaltered. Alternatively, the original crust of Mercury may have been churned up like that of the Moon, but was subsequently smoothed out again by volcanic activity. Only study of actual rocks from the planet will resolve this dispute. Other areas of particular interest for first-hand study are deep craters near the poles whose interiors are permanently shaded from the Sun, thereby preserving in deep freeze any gases that have seeped from the planet over its history.

Of all the features seen by Mariner 10, the most prominent is an

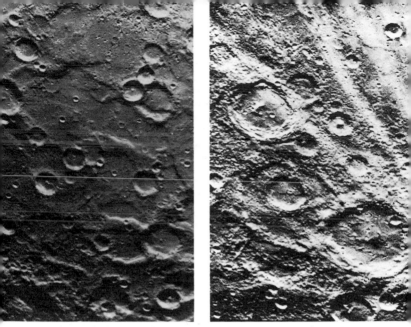

Mercury from Mariner 10. *Left:* Plains between the craters in the region of the south pole are traversed by ridges and scarp slopes. *Right:* Bright rays similar to those from craters on the Moon cross the terrain of Mercury. The largest crater seen here is 100 km in diameter. *Jet Propulsion Laboratory.*

enormous bulls-eye structure, partly hidden by shadow, named the Caloris Basin. Its diameter is 1400 km, similar to Mare Imbrium on the Moon; it was presumably formed by the impact of an asteroid after most of the surface had already been cratered. The Caloris Basin contains several concentric rings of mountains, and is surrounded by a number of radiating ridges and grooves. Most importantly from a geological point of view, its interior, and much of the low-lying land around it, has been flooded by lava. Geological activity died out on Mercury over 3000 million years ago, as it did on the Moon. Since then, little has changed except for the random arrival of a stray meteorite.

Despite its outward resemblance to the Moon, inwardly Mercury is believed to be more like the Earth. Mercury has a relatively large mass for its small diameter, which implies that it has a large iron core three quarters its diameter. A core that size would be as big as the Moon. The existence of an iron core was directly confirmed when Mariner 10 measured a magnetic field around the planet, albeit with only one per cent the strength of the Earth's magnetic field – but that is still far stronger than the magnetic fields of Venus or Mars. In its own way, Mercury turns out to be a fascinating world containing many lessons about the origin and development of the Solar System.

Venus

Many people have seen Venus without realizing it. It appears as the brilliant evening or morning 'star', the most prominent object in the twilight, unmistakably outshining every genuine star with its cold white light. So striking is Venus at its best that it is frequently reported as a hovering UFO.

Venus orbits the Sun every 225 days at a distance of 108 million km; it can pass within 40 million km of the Earth, closer than any other planet. In size (only 650 km less than the Earth's diameter) it is almost a twin of the Earth, being 12,100 km in diameter. But the brilliance of Venus in the sky is due not just to its proximity and size. The main reason is its cloak of unbroken clouds that reflect 76 per cent of the light hitting them. Although making Venus so prominent, the existence of these clouds has prevented astronomers from ever seeing its surface.

A simple pair of binoculars, or a small telescope, will show the crescent Venus as well as this photograph. Unbroken clouds veil the surface of the planet from the view of even the largest telescopes. *Hale Observatories.*

Through a telescope, Venus appears as a white billiard ball that goes through phases similar to the Moon's as it orbits the Sun. As with the Moon, one complete cycle of phases (the synodic period) takes longer than one orbit (the sidereal period) because the Earth and Venus are moving relative to each other in their respective orbits around the Sun; for Venus, the synodic period is 584 days. When it is a crescent, Venus is close enough to the Earth for its phase to be picked out in modest binoculars; some people even claim to have seen Venus as a crescent with the naked eye. The planet appears most brilliant when 28 per cent of its disk is illuminated as seen from Earth, this being the most favourable combination of distance and phase. Venus can reach a maximum magnitude of -4.4, nearly seven times brighter than the next most prominent planet, Jupiter. Being so bright, Venus is best observed against a twilight sky to reduce the dazzle.

Only the vaguest markings can be made out in the clouds of Venus with Earth-based telescopes: some dusky shadings are visible on the disk, and often the clouds appear to be brighter at the poles with a surrounding darker collar. The terminator (the edge of the illuminated portion) can appear irregular, due not so much to differences in the height of the clouds as to differences in brightness. These effects are caused by the corkscrew circulation of clouds around Venus from its equator to its pole; this was only realized from space probe photographs.

Unable to see the planet's surface because of the enveloping clouds, astronomers could only guess at the rotation period of Venus – and they guessed incorrectly. As in the case of Mercury, radar observations provided the surprising truth. It turns out that Venus rotates on its axis from east to west, the opposite direction from the Earth and other planets, and it does so very slowly – once every 243 days, longer than the 225 days it takes to orbit the Sun. Its clouds, though, rotate every four days, also retrograde (east to west), a result of high-speed winds in the upper atmosphere.

Before the days of space probes, theories abounded about the nature of the planet's surface. Since Venus was so similar in size to the Earth, it was tempting to speculate that conditions there might be Earthlike. One charming notion was that Venus resembled our planet as it had been in Carboniferous times, with steaming jungles and even dinosaurs. Some astronomers proposed that the planet was entirely covered with water, while others imagined it to be a world of deserts. None of the theories came close to anticipating the uniquely hostile conditions on Venus.

Radio astronomers provided the first clue in the late 1950s when they detected radio wave emissions from the planet which implied that it was very hot (even hotter than boiling water). The deserts of the Earth are mildly warm by comparison. These readings, doubted at the time,

The clouds of Venus circulate once around the planet every 4 days, spiralling from equator to pole as they do so. The cloud patterns are visible only in ultra-violet light, as in this photograph from the Pioneer Venus orbiter. *NASA*.

were confirmed in 1962 by the American probe Mariner 2 which scanned the planet as it flew past at a distance of 35,000 km.

Conditions on Venus were experienced directly for the first time by a Soviet probe, Venus 4, when it parachuted into the atmosphere in October 1967. It found that Venus' atmosphere is made almost entirely of unbreathable carbon dioxide gas, but the probe was destroyed by the intense heat and crushing pressure long before it reached the surface. Venus 7 was the first probe to land intact on the surface, on December 15, 1970. It registered a temperature of 475°C and an atmospheric pressure 90 times that on Earth. Venus 7 landed on the night side of the planet; in 1972 Venus 8 landed on the day side, finding conditions there to be identical. The dense atmosphere of Venus traps heat like a blanket, keeping the temperature constant over the entire planet. Under such pressure cooker conditions, the atmosphere of Venus behaves more like a liquid than a gas. To explore Venus is like trying to explore a scalding hot ocean, and it requires a refrigerated spacecraft reinforced like a submarine.

Why should Venus be so hot – hotter even than the Sunward face of Mercury, despite the fact that its clouds reflect over three quarters of the incoming sunlight? The answer lies with what is termed the greenhouse effect. About one per cent of the incoming sunlight penetrates to the planet's surface, according to the Venus 8 probe, so that it is as gloomy there as on a heavily overcast day on Earth. That incoming sunlight is absorbed by the surface and is reradiated at longer wavelengths, in the infrared. Although the carbon dioxide of the atmosphere is transparent to visible light, it traps infrared; since infrared is heat energy, the temperature of the atmosphere rises.

It turns out that Venus and the Earth have similar amounts of carbon dioxide, but on Earth most of it is locked away in rocks such as limestone. Whereas the amounts of carbon dioxide are similar, Venus has far less water than the Earth. Whatever water it originally possessed has long since evaporated and been lost to space. Only a trace of water vapour remains, but that is sufficient to boost the effect of the carbon dioxide in causing the greenhouse effect of the atmosphere of Venus.

A final contribution to the greenhouse effect is provided by the clouds of Venus. These are made not of water vapour, as are the clouds of Earth, but of sulphuric acid of 80 per cent concentration, stronger than in a car battery. Sulphuric acid, too, absorbs infrared. Taken together, these three factors of carbon dioxide, water vapour and sulphuric acid turn Venus into a perfect Suntrap. The clouds add to the nastiness of Venus in another way: from them descend showers of corrosive sulphuric acid rain. Despite its heavenly name, Venus turns out to be an incarnation of Hell.

In December 1978 a flotilla of five American space probes called Pioneer plummeted into the atmosphere of Venus. They found that the uppermost layer of sulphuric acid clouds, the one which observers view through telescopes, lies about 65 km above the planet's surface and is a few kilometres thick. Around 58 km altitude is a thin haze layer apparently consisting of sulphuric acid particles which give the clouds a yellow tinge. The densest cloud layer of all occurs at about 50 km altitude, and it is from this layer that the rain of sulphuric acid droplets falls. Below the clouds, the gloom is broken by flashes of lightning which strike so frequently they merge into a continuous glow, while thunder reverberates in the atmosphere.

Although the clouds of Venus mask its surface from view, astronomers have nevertheless been able to map the planet's features thanks to the fact that radio waves penetrate the clouds. Radar observations from Earth during the 1970s revealed some features, but the first complete map of the planet was made by a radar device aboard a Pioneer spacecraft that went into orbit around Venus in December 1978, a companion to the probes that entered the atmosphere.

Venus is mostly rolling plains, but there are three main continental areas. One, called Ishtar Terra, the size of the United States, has mountains which tower 12 km above mean surface level, higher than Mount Everest on Earth. The largest continental area of all, Aphrodite Terra, the size of South America, is cut by a system of rift valleys that extend for thousands of kilometres. The highlands of Venus are believed to have been formed by volcanic action, and those volcanoes may still be erupting today, triggering the lightning that seems to concentrate around the highlands.

Photographs from Soviet lander probes sitting on the surface of Venus show a rocky wasteland bathed in a sulphurous orange glow. Chemical analyses by these probes confirm that the surface rocks of Venus are similar in composition to volcanic basalts on Earth, as would be expected from the volcanic activity on the planet. Despite the information from these probes, there are many questions that remain to be answered about the past evolution of the planet Venus. Venus remains shrouded in mystery, a tantalizing vision of an Earth that might have been – and a terrifying demonstration of what the Earth itself might have become had it been born closer to the Sun.

Mars

Mars is distinguishable by its intense reddish-orange hue, stronger than the colour of any star and the cause of its association with the God of War. At its best Mars can shine at magnitude −2, rivalling Jupiter. But the most favourable appearances of the planet are few and far between, the main reason being the marked ellipticity of its orbit which takes it between 206–249 million km from the Sun (average distance 228 million km). If the Earth passes Mars when Mars is at its closest to the Sun, only about 57 million km separates the two bodies and astronomers get their best views of the red planet. But when Mars is farthest from the Sun, nearly 100 million km separate it from Earth, and it appears unimpressive even in powerful telescopes. The closest approaches of

Four faces of Mars, photographed as the planet rotated. Top left, Syrtis Major is centrally placed. Bottom right, Mare Acidalium is the large dark area in the north with Mare Erythracum in the south *Lunar and Planetary Laboratory*.

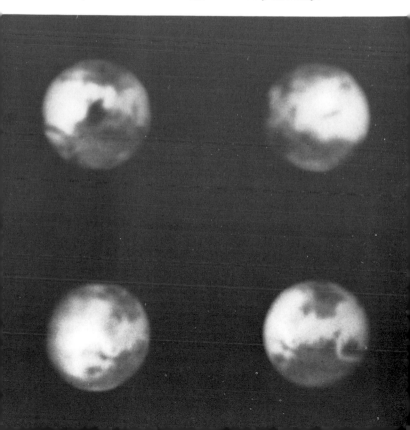

Mars occur at intervals of about 15 years, so astronomers do not waste their chances of observing Mars at its best.

Mars has a diameter of 6790 km, just over half that of the Earth. Its day is just over half an hour longer than our own – 24 hours 37 minutes – but its year is nearly twice as long as ours, 687 Earth days. Its orbit lies outside that of Earth, so Mars can never appear as a crescent. But at times it displays a distinctly gibbous phase, like the Moon when it is a couple of days from full.

Binoculars show Mars as nothing more than an orange dot of light; a telescope is needed to bring into view the main features of the planet. Among the most obvious are the white polar caps, which stand out in stark contrast to its ochre-coloured deserts. Occasionally, violent winds whip up dust storms in the thin atmosphere, obscuring the dusky surface markings. Frustratingly for observers, the worst dust storms tend to occur when Mars is at its closest to the Sun, and thus ruin the best observing opportunities. The most prominent dark surface marking, a large triangular area named Syrtis Major, was first noted by the Dutchman Christiaan Huygens in 1659. Syrtis Major, along with the polar caps, should be visible in a modest amateur telescope.

The temptation to assume similarities between Mars and the Earth led early astronomers astray. The dark areas, which range in colour from brown to grey-green, were termed seas and lakes in the belief that they really were filled with water, while the orange areas were named after places on Earth – there is an Arabia, Libya, Syria and Sinai on Mars. Towards the end of the 19th Century astronomers realized that there were no oceans on Mars after all, but this opened the way for a much more intriguing explanation for the dark areas: that they were covered with primitive vegetation, such as moss or lichen. In support of this view, observers noted that when the polar caps melted in the Martian summer, the surface markings became larger and darker; this was interpreted as being due to the vegetation growing in the milder, wetter conditions.

The most extreme proponent of the life-on-Mars idea was an American astronomer, Percival Lowell. His ideas were inspired by the discovery made in 1877 by an Italian, Giovanni Schiaparelli, of apparent long, straight lines crisscrossing the planet's surface. Schiaparelli called these *canali*, a word which strictly means 'channels'; but inevitably it was translated as 'canals', implying they were artificial, although Schiaparelli himself kept an open mind about their true nature.

For Lowell, there was no doubt: the canals were evidence of an advanced civilization on Mars. His beliefs sparked off a whole generation of science fiction, including the famous H. G. Wells *War of the Worlds*. Lowell set up his own observatory at Flagstaff, Arizona, in 1894 specifically to study Mars. There, he produced fanciful maps of the canal network and wrote books such as *Mars as the Abode of Life*, pub-

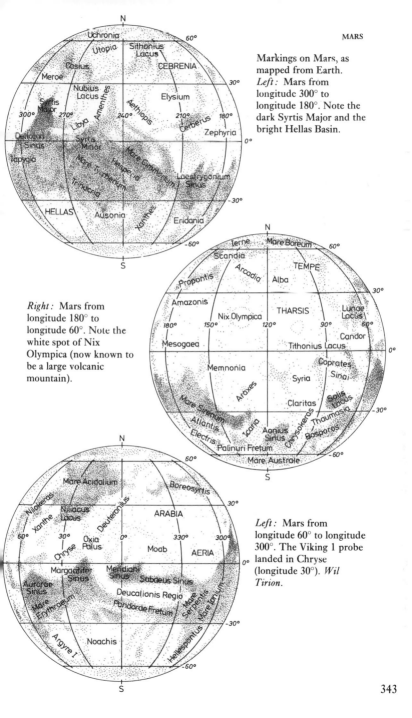

Markings on Mars, as mapped from Earth. *Left:* Mars from longitude 300° to longitude 180°. Note the dark Syrtis Major and the bright Hellas Basin.

Right: Mars from longitude 180° to longitude 60°. Note the white spot of Nix Olympica (now known to be a large volcanic mountain).

Left: Mars from longitude 60° to longitude 300°. The Viking 1 probe landed in Chryse (longitude 30°). *Wil Tirion.*

343

lished in 1908, in which he expounded his theory of a Martian civiliz-ation clinging to existence on an arid planet, reliant on the canals to bring melt water from the polar caps to irrigate their crops at the equator.

Most other astronomers failed to see the canals or in their place could detect only broad, irregular smudges. After Lowell's death in 1916 a few devoted followers kept the canal theory going, but the idea of Martian civilization was doomed to oblivion in the light of new know-ledge about conditions on the planet.

By the 1950s it was clear that the atmosphere on Mars was far too thin for a human to breathe. Under so thin an atmosphere, temperatures would be frigid and dangerous amounts of ultraviolet radiation would penetrate to the ground. To make matters worse, no oxygen could be detected in the planet's atmosphere, although carbon dioxide was known to be present. No advanced lifeform could exist under such conditions, although hardy vegetation still wasn't ruled out.

Our knowledge of Mars took a major step forward in July 1965 when the American space probe Mariner 4 sent back the first close-up photo-graphs as it flew past Mars at a distance of 10,000 km. Its most important revelation was that there are craters on Mars, looking similar to those on the Moon, only somewhat more eroded because of the effects of the planet's atmosphere. Actually, the discovery of craters on Mars was not unprecedented. The great American observer E. E. Barnard saw them in 1892 with the 91 cm (36 inch) refracting telescope at Lick Observ-atory, as did J. E. Mellish in 1915 using the Yerkes Observatory's 102 cm (40 inch) refractor, but neither man published his observations for fear that they would be disbelieved.

Unfortunately, a lunar-like Mars seemed an unpromising abode for life of even the lowliest form. The image of Mars as a dead world, in both the geological and biological sense, was reinforced in 1969 when Mariners 6 and 7 photographed more craters apparently of impact origin. The first full survey of the entire planet did not come until 1971–2 when Mariner 9 went into orbit around Mars. It revealed major forma-tions that previous Mariners had, by ill luck, completely missed.

For a start, there was a chain of three volcanic mountains topping a highland area known as the Tharsis Bulge; they are named Arsia Mons, Pavonis Mons and Ascraeus Mons. (The word Mons means mountain.) Other types of feature on Mars and the names given to them include lowland plains (Planitia); highland plateaus (Planum); valleys (Vallis); canyons (Chasma); and eroded craters (Patera). Northwest of the chain of Tharsis volcanoes is an even bigger volcanic mountain, Olympus Mons, originally seen from Earth as a white ring and then known as Nix Olympica ('the snows of Olympus'). This whole area is noted for the frequent appearance of white clouds, often described as being shaped like a W. The existence of mountains here explains the preference of clouds for this region.

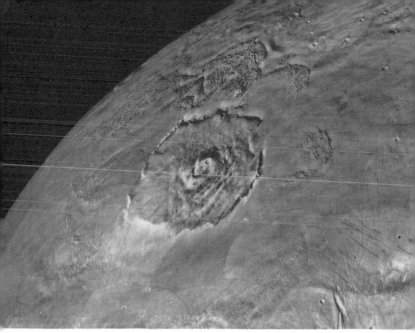

Olympus Mons, an immense volcanic mountain, blisters the surface of Mars in this mosaic of photographs from the Mariner 9 space probe. Ancient lava flows spread out around it. Olympus Mons is visible to observers on Earth as a white ring, long known by the name Nix Olympica. *Jet Propulsion Laboratory*.

Olympus Mons, 600 km wide and 26 km high, is larger even than the volcanic islands of Hawaii on Earth; it is often given the title of 'largest volcano in the Solar System', although there may be larger ones on Venus. Whatever the case, Mars certainly has impressive volcanoes.

Another dramatic and unexpected feature was an immense system of faults, 75 km wide and up to 7 km deep, extending east from the Tharsis Bulge. This massive rift valley, now called the Mariner Valley, not merely dwarfs the Earth's Grand Canyon: at 4000 km long it could span the entire United States. The Mariner Valley corresponds to a broad, fuzzy canal seen from Earth, called Coprates; its visibility from Earth is mainly due to the fact that dark dust collects along its floor. Coprates is one of the few 'canals' that has any corresponding surface feature on Mars. A few other canals, notably one called Cerberus, do coincide with dark streaks on the Martian surface, caused by darker dust or rock; but these are all broad and irregular markings, unlike the thin, straight canals that Lowell drew. Close comparison of Lowell's canal maps with space probe pictures reveals few correlations at all. There seems no real explanation for the canal network that Lowell and his followers drew, other than the fallibility of human observers straining at, or past, the limits of visibility.

Although the Lowellian canals were not in evidence, Mariner 9 found convincing signs that water had once flowed on Mars. Sinuous channels, looking like dried up river beds, snaked across parts of the planet's surface. Some of the lowlands appeared to have once been inundated by flash floods. Liquid water cannot exist on Mars today because the atmospheric pressure is too low. The existence of ancient water courses implied that in the past the atmosphere was denser – and a denser atmosphere would keep the planet warmer. Perhaps the volcanoes of Mars gushed out enough gas to change the climate temporarily. If that were so, life might have had a chance to arise on Mars after all, and microscopic organisms such as bacteria might still cling to existence among the red sands of the planet. The only way to find out was to go down and look.

In 1976 two American space probes called Viking were sent to look for life on Mars. Each probe came in two halves: a lander and an orbiter. The Viking 1 lander touched down on a lowland plain called Chryse in the northern hemisphere, over which water appeared to have flowed during the wet times on Mars. The other lander descended on the opposite side of the planet in an area known as Utopia which is encroached upon by the outer fringes of the north polar cap during winter.

The landers carried colour cameras as well as instruments to analyse the soil and the atmosphere. Each Viking lander looked around, finding itself on a rock-strewn desert, without any visible signs of life – no plants, insects or animal tracks. A mechanical arm reached out to pick up samples of the soil and tip them into an on-board biological laboratory which then set to work in search of Martian microorganisms.

To the disappointment of many, neither Viking found life in the soil of Mars, despite exhaustive tests. That is not to say that Mars is necessarily totally sterile; maybe there is life of a kind that would not respond to the Viking experiments, or possibly it is concentrated in certain oases that the Vikings missed. But the negative Viking results do make it extremely unlikely that there is, or ever has been, anything alive on Mars.

A lifeless Mars is not too surprising in the light of readings from Viking's other instruments, which revealed just how hostile conditions on the planet really are. The Viking 1 lander recorded a maximum summer afternoon air temperature of $-29°C$, whereas farther north at the Viking 2 landing site temperatures dropped well below $-100°C$ as winter approached and the carbon dioxide in the atmosphere began to freeze, forming patches of white frost on the surface. Atmospheric pressure was a mere 7.5 millibars (750 Pascals), equivalent to the pressure at a height of 35 km above the Earth.

Although Mars was a disappointment for the biologists, there was

Rusty-red surface of Mars photographed by the Viking 2 lander. Part of the lander itself is visible in the foreground. The largest rocks are about 1 m across. Fine dust suspended in the atmosphere makes the sky pink. *NASA*.

plenty to interest the geologists. As expected, the rusty redness of Mars is caused by large amounts of iron oxide in the surface rocks. Mars may be the richest source of iron ore in the Solar System. Even the sky is pink, from fine particles of dust suspended in the atmosphere.

While the landers studied Mars from the surface, the orbiters continued the surveying job begun by Mariner 9. Mars turns out to be a planet of two differing halves. The northern hemisphere of the planet is the lower and smoother of the two, flooded by lavas from the volcanoes of Mars. The southern hemisphere is higher and has been heavily cratered by meteorite impacts, giving it a broad similarity to the lunar highlands. This was the area that the first Mariners photographed. In the southern hemisphere of Mars are two large impact basins, Hellas and Argyre. Argyre, 1000 km in diameter, is about the same size as the Moon's Mare Imbrium, while Hellas is even larger, equivalent in size to Oceanus Procellarum. Unlike the lunar maria, however, Argyre and Hellas appear to be filled with light-coloured dust.

It comes as a surprise to learn that Mars is not short of water, although most of that water exists in frozen form, in the polar caps and in a subsurface permafrost layer. This offers another possible explanation for the ancient water channels of Mars: the melting of subsurface ice by volcanic heating. The polar caps of Mars are made of

Phobos, the larger of the two moons of Mars, is an irregularly shaped lump of rock. The crater at its north pole is 5 km across; the smallest craters visible in this Viking picture of Phobos are a few hundred metres wide. *NASA.*

water ice a few metres thick, augmented each winter by carbon dioxide. This freezes out of the atmosphere to produce a thin sprinkling of frost that can extend more than halfway to the equator. During each Martian year the atmospheric pressure changes by about 20 per cent as carbon dioxide evaporates from one polar cap with the coming of spring, migrates to the opposite hemisphere and then freezes out again as winter arrives at the other pole.

There is one final puzzle from the pre-space probe era of observation that remains to be explained: what caused the seasonal changes in the dark areas if there is no vegetation on Mars? Wind-blown dust provides the answer. Mariner 9 and the Viking orbiters recorded many examples of changes in surface markings caused by light and dark dust being blown by seasonal winds. The Syrtis Major, for example, is a sloping area of darker rock which periodically becomes partly covered by paler dust and is later swept clear again. Dust blown by the winds on Mars, which can reach 200 km per hour, is also a powerful erosive agent, sandblasting the surface features of the planet.

No discussion of Mars would be complete without reference to its two tiny moons, Phobos and Deimos. They were discovered in 1877 by Asaph Hall using the 66 cm (26 inch) refractor at the U.S. Naval Observatory in Washington, D.C., and both are beyond the range of normal amateur telescopes. Space probe photographs show that both are cratered lumps of rock shaped like lumpy potatoes. Phobos, larger and closer to Mars, measures about 27×19 km; Deimos is about 15×11 km.

Phobos is remarkable in that is orbits Mars three times a day. It is the only known moon in the Solar System that orbits more quickly than its parent planet spins. It also lies closer than any other moon to its parent – a mere 6000 km above the surface of Mars. Astronomers are unsure whether Phobos and Deimos formed in orbit around Mars, or whether they are asteroids which strayed too close to Mars and were captured by the planet's gravity. Whichever is the case, they will make a fascinating sight in the sky for the first astronauts that set foot on Mars, the planet of red, frozen deserts which was so nearly suitable for life of its own.

Jupiter

Jupiter is the king of the planets, and the most fascinating of all to study with small instruments. Humble binoculars reveal the planet's cream-coloured disk and four main moons, called the Galilean satellites after Galileo who discovered them in 1609. Individuals with exceptionally acute vision claim to be able to see the Galilean satellites with the naked eye.

A small telescope brings into view some of the details of Jupiter's disk: dark belts of cloud parallel to the equator and an eye-shaped marking in the southern hemisphere known as the Great Red Spot, first observed in 1665. Careful study of these features reveals that Jupiter's rotation period varies with latitude, from 9 hours 50 minutes at the equator (the fastest of any planet in the Solar System) to 9 hours 55 minutes at higher latitudes. In addition, the Great Red Spot drifts slightly in relation to its surroundings. These effects demonstrate that the visible surface of Jupiter is not solid. We are looking at clouds, constantly seething and swirling, changing in colour and shape. Jupiter never appears the same twice. That is its attraction.

The giant planet Jupiter looms large in this photograph taken from Voyager 1 in February 1979. At bottom left is the Great Red Spot; to its right, in front of the planet's disc, is Jupiter's orange moon Io. *NASA.*

350

Jupiter's Great Red Spot stares like a Cyclopean eye from this Voyager 2 photograph taken in July 1979. Other clouds exhibit fantastic scalloping as they sweep around the Red Spot and a white oval to the south of it. *NASA.*

Jupiter is easy to spot with the naked eye, of magnitude −2.5 when at its closest to Earth, 588 million km away. And it manages to outshine every star except Sirius even when at its most distant from us. Its brightness results from its highly reflective clouds and its imposing size, the largest planet in the Solar System. Examination of its outline gives added confirmation to the fact that it is not a solid planet: Jupiter has a bulging midriff. The diameter at its equator is 142,800 km, but from pole to pole it measures 134,200 km. A line of 11 Earths would be needed to equal Jupiter's equatorial width. And even though Jupiter is made of gas, its bulk is such that its mass is $2\frac{1}{2}$ times as much as all the other planets together.

Jupiter orbits the Sun every 11.9 years at an average distance of 778 million km, over five times the distance of the Earth from the Sun. It is well placed for observation from Earth every 13 months. Because Jupiter's cloud features are so impermanent and mobile, it is impossible to give more than a generalized description of the planet's appearance. Its disk is crossed by alternating bright zones, caused by ascending gas, and dark belts where the gases descend. Frozen ammonia crystals form high, cold clouds in the bright zones; the darker belts are lower and warmer ('warm' is only a relative term, for the average temperature of the cloud tops is $-150°C$). The colours of the belts can vary from yellow and brown to orange, red or even purple as a result of complex chemicals in the atmosphere of Jupiter.

High-speed winds up to 400 km per hour whip the edges of the zones and belts into turbulent eddies, giving them a scalloped appearance. The weather on Jupiter is unpredictable. Dark and light spots can suddenly erupt in the clouds, lasting for weeks or even decades before fading away. One of the main roles of amateur observers is to track these storms as they erupt and move around the planet.

Of all the markings on Jupiter, the most famous, and by far the most permanent, is the Great Red Spot. It certainly is great: 14,000 km wide and 40,000 km long, enough to swallow three Earths. But it is not always red; most often it is pinkish, and sometimes it can fade to a colourless grey. Its colour is believed to come from either red phosphorus or sulphur. By good fortune, the spot was particularly prominent when the Voyager 1 and 2 spacecraft reached the planet in 1979. Even now its nature is not fully understood, but it appears to be an upward spiralling column of gas similar to a hurricane on Earth, its top spreading out about 8 km above the surrounding cloud deck. Jupiter's other, smaller spots are believed to be similar swirling storm systems. As if to emphasize the storminess of the planet, the Voyager probes photographed massive flashes of lightning on Jupiter's night side.

The key to Jupiter's meteorology is the fact that it gives off twice as much heat as it receives from the Sun. The planet was hot when it formed, and still retains some of that heat today. This internal store of heat drives the complex cloud systems of Jupiter, keeping the Great Red Spot and its smaller relatives alive for far longer than any storms on Earth.

Interestingly, Jupiter has virtually the same chemical composition as the Sun: mostly hydrogen and helium. There is thought to be a rocky core about twice the size of the Earth at the centre of Jupiter, but no space probe could ever land on it. Beneath the wispy high-altitude clouds of frozen ammonia are complex chemicals that give the dark belts their colour. Deeper still, temperatures are similar to those on Earth; here are clouds of water vapour. About 1000 km below the visible cloud tops, temperatures and pressures have increased to the point where hydrogen is compressed into a liquid. The liquid hydrogen seas of Jupiter are about 20,000 km deep. Below, under the crushing pressure of three million Earth atmospheres, hydrogen is compressed into a super-dense state with the properties of a metal; hence it is known as metallic hydrogen. Convection within the hot metallic hydrogen interior of Jupiter is thought to be responsible for the planet's intense magnetic field, 10 times stronger than the Earth's, extending 100 times Jupiter's radius into space. If the magnetic field around Jupiter were visible to the naked eye, it would appear over twice the size of the full Moon.

Jupiter is the centre of a fascinating collection of satellites, like a mini Solar System. We have already mentioned the four largest, known as the Galilean satellites. With the simplest optical aid they can be seen performing a merry dance around Jupiter – changing position from night to night, sometimes out of sight behind the planet, sometimes transiting across its face, and sometimes being eclipsed in its shadow.

The closest of the Galilean satellites to Jupiter is Io, 3600 km in diameter (slightly larger than our own Moon), orbiting every $42\frac{1}{2}$ hours. Io is the most volcanically active body in the Solar System. The Voyager 1 probe in 1979 photographed eight volcanoes erupting simultaneously on it. Hundreds of other volcanic vents were visible, though not actually erupting. Those volcanoes erupt not lava but liquid sulphur which solidifies to form the red, orange and yellow of Io's surface.

What keeps Io molten remains an open question. According to one theory, Io is caught in a gravitational tug of war between Jupiter and the other Galilean satellites; their opposing pulls release tidal energy that melts Io's interior. But the amount of energy generated by this mechanism may not be sufficient, and the answer may instead lie with powerful electrical currents that flow through Io as it orbits within Jupiter's magnetic field, heating the moon like an electric bar fire.

Io recycles its interior onto its surface, endlessly turning itself inside out. Some of the sulphur escapes and showers onto the innermost moon of Jupiter, Amalthea, giving it an orange coating. Amalthea is an irregularly shaped lump of rock only about 200 km in diameter, too faint to be seen in amateur telescopes.

Within the orbit of Amalthea the Voyager probes discovered a faint ring of dust, a mere 50,000 km above Jupiter's cloud tops. This tenuous ring of Jupiter is believed to result from the break-up of one or more tiny moons, also discovered by the Voyagers, which orbit at the ring's outer edge.

Moving outwards past Io we come to Europa, smallest of the Galilean satellites with a diameter of 3100 km. Europa is encased in a white shell of ice, veined with fine cracks. Underneath its ice crust, Europa is believed to have a rocky interior. Next in line from Jupiter is Ganymede, the largest and brightest of the satellites; what's more, at 5200 km in diameter it is the largest moon in the Solar System, larger even than the planet Mercury. Ganymede and the fourth of the Galilean satellites, Callisto, 4800 km in diameter, are both balls of rock and ice, rather like giant muddy snowballs. Callisto is saturated with impact craters, the largest being 300 km in diameter and called Valhalla; it is similar to the large basins on the Moon and Mercury, surrounded by wave-like ridges. Ganymede is also cratered by impacts, but less heavily than Callisto. It also exhibits a strange grooving on its surface. Bright patches

A volcano called Loki erupts at the limb of Jupiter's moon Io, sending clouds of sulphur 100 km into space; photographed by Voyager 1 in March 1979. *NASA*

on Ganymede and Callisto exist where recent impacts have exposed fresh ice.

The rest of Jupiter's total of at least 16 moons are small and insignificant. Several of these, particularly the outer four which move in highly elliptical, retrograde orbits, may be passing bodies that were captured by Jupiter's gravity.

Jupiter's orange moon Io.

Europa has a cracked, icy surface.

Ganymede, largest moon of Jupiter.

Callisto is dark and cratered. *NASA.*

Saturn

Bright rings girdle Saturn's equator, making it the most beautiful of the planets. Those distinctive rings can be spotted clearly in a small telescope; good binoculars, mounted steadily, will show the small disk of the planet elongated into an ellipse by the rings. Binoculars should also pick out Saturn's largest moon, Titan, which orbits the planet every 16 days.

Saturn's rings can reflect more light than the body of the planet itself, so that at its best Saturn appears of magnitude -0.3. Without the rings, Saturn would be no brighter than magnitude 0.7. Strangely enough, from time to time Saturn can appear to be without rings. The reason is that the planet's axis is tilted at 29° to the vertical. As Saturn orbits the Sun, sometimes the rings are tipped towards us, while at other times the rings are presented edge-on. So thin are the rings that when edge on (as happens about every 15 years) they disappear from view in even the largest telescopes on Earth. The brightness of Saturn therefore depends not only on its distance from Earth but also on the aspect of the rings.

Saturn and its rings presented a serenely beautiful sight to the Voyager 1 space probe as it approached the planet on October 30, 1980. Note Cassini's division in the rings, and the shadow of the rings on Saturn. *NASA*.

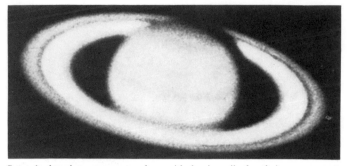

Saturn's changing appearance: *above*, with the rings tilted at their greatest angle towards us; *below*, with the rings edge-on, when they disappear from the view of observers on Earth. *Lowell Observatory.*

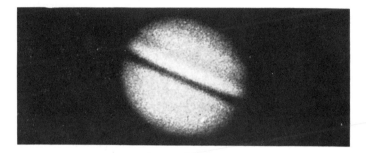

The ringed planet orbits the Sun every $29\frac{1}{2}$ years at an average distance of 1430 million km, $9\frac{1}{2}$ times farther than the Earth is from the Sun. In many ways, Saturn is a smaller brother of Jupiter. Its equatorial diameter of 120,000 km is second only to that of Jupiter; its rotation period of $10\frac{1}{4}$ hours is second fastest to that of Jupiter; and Saturn is another planet made mostly of hydrogen and helium gas.

In one way, though, it is unique among the planets: its average density is less than that of water. This remarkable fact comes about because the planet's mass is less than a third of Jupiter's mass, so its gravity is less and hence its central regions are not compressed as densely. There probably is a rocky central core, but the region surrounding the core, in which hydrogen is compressed into a liquid metallic form, extends out to only half the planet's radius, against three quarters of the radius in the case of Jupiter. That is not sufficient to compensate for the low density of Saturn's outer layers, so the planet's overall density is a mere 70 per cent of that of water. Given a big enough ocean, Saturn would float on it.

Saturn's low density is apparent in another way – its outline, which is even more squashed than that of Jupiter. Saturn's pole-to-pole diameter is 108,000 km, fully 10 per cent less than its equatorial diameter; on Jupiter the difference is 6 per cent.

Through a telescope, Saturn appears as a tranquil, ochre-coloured disk, darker at the poles and with some dusky horizontal bands. There are none of the swirling, multicoloured storm clouds that make Jupiter's globe so interesting in a small telescope, and certainly nothing resembling the Great Red Spot. Serious study of Saturn requires telescopes at least 20 cm in aperture.

This is not to say that Saturn lacks activity in its atmosphere; it is just that the cloud patterns are concealed by high-altitude haze. Cameras aboard the probes Voyager 1 and 2 which reached Saturn in 1980 and 1981 recorded low-contrast cloud swirls resembling those on Jupiter. The meteorology of both planets should be similar, for they both have internal sources of heat. Saturn radiates $2\frac{1}{2}$ times as much heat as it receives from the Sun, a legacy of its birth. Tracking of cloud systems revealed that gales of up to 1800 km per hour blow on Saturn, four times faster than on Jupiter.

Inevitably, the attention of the space probes focused on the glorious rings. As seen from Earth, the rings look like a continuous disk encircling the planet, but appearances are deceptive. The Dutchman Christiaan Huygens in 1655 was the first to realize the rings are not solid, but are in fact made up of a swarm of tiny particles orbiting Saturn. The central part of the ring, known as ring B, is the widest and brightest. It is separated from the outer, fainter ring A by a 3000 km-wide gap known as Cassini's division. Extending inwards from ring B towards the planet is the faintest ring of all, the transparent ring C, also known as the crepe ring. A narrow gap called Encke's division can also be seen in the A ring, while observers looking through large telescopes under good conditions reported apparent ripples within the rings, a sign that the density of ring material varies from place to place.

But even the best telescopic views could not have prepared astronomers for the astounding wealth of detail that was revealed by space probes. Under the close scrutiny of the Voyagers' cameras, the rings broke up into thousands of narrow ringlets and gaps, like the ridges and grooves of a gramophone record. Some ringlets were not perfectly circular but were elliptical in shape. Even the Cassini division was not empty, but contained threadlike ringlets. A new outer ring, called the F ring, appeared to consist of twisted strands like a rope.

In some places, fine dust overlays the rings, apparently supported by electromagnetic forces in Saturn's magnetosphere, producing darker features known as spokes. Spoke-like features have been reported from time to time by ground-based observers, but it took the Voyager pictures to establish their reality.

In close-up, Saturn's rings break up into thousands of ringlets like the ridges and grooves of a gramophone record. The colours have been artificially enhanced by computer. Voyager 2 photograph, taken August 17, 1981. *NASA*.

The particles that make up Saturn's rings range in size from tiny specks to lumps the size of a house or larger. Their composition is mostly frozen water, possibly mixed with dust, resembling loosely compacted snowballs. The rings are most likely material that was prevented from forming into a moon by the overpowering force of Saturn's gravity. Alternatively, they are the remains of a former moon that strayed too close to the planet and broke up.

Saturn's rings are extraordinarily thin in relation to their 270,000 km diameter. Voyager observations showed that the rings are no more than 100 m thick. On the same ratio of thickness to diameter, a gramophone record would be 5 km in diameter.

The Voyagers discovered several moons of Saturn too small to be seen from Earth, bringing the total of known Saturnian moons to over 20. One of these tiny moons, known simply as S15, patrols the outer edge of ring A. Two others, S 13 and S 14, orbit either side of the F ring, shepherding its particles. Saturn has nine satellites with names: Mimas, Enceladus, Tethys, Dione, Rhea, Titan, Hyperion, Iapetus and Phoebe. (A tenth satellite, announced in 1966 and named Janus, turned out to be a misidentification of two separate satellites, now called S 10 and S 11, moving along the same orbit.)

The gravitational effect of some of these satellites helps produce the gaps in Saturn's rings. For instance, the gravitational tug of Mimas pulls particles out of the Cassini division. Mimas itself is, like most of the other moons of Saturn, a dirty snowball of frozen water and rock.

It sports a remarkable crater 130 km in diameter, larger than Copernicus on the Moon, fully a third of its own 390 km diameter. The impact that caused this grotesque feature must have nearly shattered Mimas.

Iapetus, Saturn's outermost moon but one, is another strange body: one side is five times darker than the other. This harlequin effect is probably caused by dust knocked off Saturn's outermost moon, Phoebe, which is the darkest of Saturn's moons. The dark dust from Phoebe falls inwards to be swept up by Iapetus, coating its leading face while bright ice on its trailing side remains exposed.

Saturn's largest moon, Titan, 5120 km in diameter, deserves to be considered as a planet in its own right. Bigger than Mercury (but slightly smaller than Jupiter's main moon Ganymede) it is the only moon to have a substantial atmosphere – in fact, the atmospheric pressure on the surface of Titan is 50 per cent greater than it is at sea level on Earth. Titan's atmosphere is 90 per cent nitrogen, with methane making up most of the rest. Clouds of orange smog top the atmosphere, obscuring the surface from the prying eyes of space probes. Despite that thick atmosphere, Titan's surface temperature is very low, about −180°C. Perhaps a rain of liquid methane falls from the rust-coloured sky into seas of methane on Titan's surface. Titan has been described as a deepfreeze version of the Earth. It will certainly be a fascinating spot for a space probe of the future to land.

A massive crater, 135 km across with a central peak, scars the surface of Saturn's battered moon Mimas. The impact that produced it must have almost broken up the entire satellite. Voyager 1 photograph, November 12, 1980. *NASA*.

Uranus, Neptune and Pluto

Uranus should in theory be visible to the naked eye: at its brightest it is of magnitude 5.5. But it is so insignificant that it was never noticed by ancient astronomers, for whom the Solar System stopped at Saturn. Uranus was not discovered until March 13, 1781, when William Herschel spotted it through his telescope during a systematic survey of the skies. Once its position is known, Uranus can easily be followed in binoculars as it moves against the background stars. There is a certain fascination in seeing the blue-green disk of this distant planet through a telescope, but even in large telescopes Uranus displays no detail.

Uranus is one of the four 'gas giants' of the outer Solar System, the others being Jupiter, Saturn and Neptune. Uranus is 52,000 km in diameter at the equator, less than half that of Saturn but four times larger than that of the Earth. It lies 2900 million km from the Sun, 19 times farther than the Earth. Seasons would pass slowly on Uranus, for it takes 84 years to complete one orbit, but the seasons would be extreme, for Uranus appeared to have been knocked over onto its side: the planet's axial tilt is $98°$, meaning that its axis of rotation lies almost in the plane of its orbit. Every 42 years, therefore, one of the poles of Uranus is pointing towards the Sun, while its opposite pole is in darkness for decades. In between times its equatorial region faces sunwards. During each 84-year orbit of Uranus the Sun can appear overhead at every latitude, something that never happens on any other planet. No one knows why Uranus should be lying on its side in this unique way; perhaps it suffered a collision with another large body long ago.

Uranus is celebrated for another reason: it was the second planet discovered to have rings. In 1977, astronomers watched as Uranus passed in front of a star. Unexpectedly, they noticed that the star winked on and off a number of times both before and after it was obscured by the disk of Uranus. From this observation, and others like it which followed, astronomers deduced that Uranus is encircled by nine rings, so faint they are invisible by direct observation from Earth.

The rings are narrow – from 10–100 km wide, separated by gaps, from 300 km to over 2000 km wide. The rings lie 15,000–25,000 km above the cloud tops of Uranus, and are probably caused by the break-up of a former moon, or moons, that strayed too deeply into the planet's gravitational clutches. In addition to the rings, Uranus has five moons: Miranda, Ariel, Umbriel, Titania and Oberon, ranging in size from 300–1100 km in diameter. The moons, like the rings, all move around Uranus's crazily tilted equator.

In structure, Uranus is believed to have a rocky core surrounded by a mantle of ice, topped by an atmosphere of hydrogen and helium mixed with a fair proportion of methane, which gives the greenish colour. Neptune, the next-farthest planet from the Sun, is a twin of Uranus. It is slightly smaller - diameter 48,000 km – but shows the same featureless greenish disk through a telescope. Of 8th magnitude at brightest it is completely invisible to the naked eye, but can be followed in binoculars when one knows where to look.

Unlike the accidental discovery of Uranus, the existence of Neptune was predicted in advance of its discovery. Astronomers found that Uranus was not keeping to its expected course, and one suggested reason was that it was being pulled by the gravity of an as yet unseen planet. In England, mathematician John Couch Adams calculated the new planet's position in 1845, and the following year Urbain Leverrier of France came up with a similar conclusion. Neptune was found close to the predicted position by astronomers at the Berlin Observatory on September 23, 1846.

Neptune crawls around the Sun every 165 years at an average distance of 4500 million km, 30 times farther away than the Earth. Being so distant from the Sun, Neptune is a very cold and dark planet. Neptune's main distinction is its curious pair of satellites. Triton, the largest, is in a retrograde (east to west) orbit. Tidal forces from Neptune mean that Triton's orbit is gradually shrinking, so that the moon will spiral closer to the planet until it is broken up in 10–100 million years' time. A shattered Triton will form a substantial set of rings around Neptune, for it is 3700 km in diameter, larger than our own Moon. The outer satellite of Neptune, called Nereid, is much smaller, only about 500 km across, but it has an exceptionally elliptical orbit that swings it between 1.4–9.7 million km of Neptune every 360 days. Something seems to have disturbed the satellite system of Neptune; exactly what happened remains one of the mysteries of the outer Solar System.

Pluto is the odd man out among the planets – indeed, many astronomers doubt whether it deserves the title of planet at all. It is by far the smallest planet – its diameter of about 3000 km is less than that of our own Moon, and considerably smaller than Neptune's major satellite Triton. Pluto seems to be nothing more than a low density ball of frozen gas, in common with the moons of the outer planets, including Triton. One suggestion is that Pluto is a former satellite of Neptune that was somehow flung out onto a separate orbit around the Sun, and that the strange orbits of Neptune's remaining moons Triton and Nereid are consequences of that disruptive event.

Pluto certainly has the oddest orbit of any planet: it crosses the orbit of Neptune, so that at times Neptune temporarily becomes the outermost planet of the Solar System, as is the case between January 1979 and March 1999. Pluto's average distance from the Sun is 5900 million km.

Pluto is not alone in its 250 year orbit around the Sun. In 1978 astronomers discovered that Pluto has a moon, called Charon, fully 40 per cent of Pluto's diameter. Charon orbits its parent every 6.4 days, the same time that the planet takes to spin on its axis. Charon therefore hangs over one spot on the surface of Pluto, visible permanently from one hemisphere of the planet but invisible from the other.

Being so faint and insignificant, Pluto was not discovered until 1930. For decades prior to that, various astronomers had tried to predict where a planet beyond Neptune might lie, but without success. In the end, Pluto was found by Clyde Tombaugh at the Lowell Observatory in Arizona as the result of a deliberate round-the-sky photographic survey to detect new planets. Tombaugh's search failed to detect any sign of a tenth planet lurking beyond Pluto.

Comets and Meteors

Comets are insubstantial bodies, loosely knit collections of frozen gas and dust that loop through the Solar System on elongated orbits, returning to the Sun at intervals ranging from a few years to many thousands of years. A cloud of thousands of millions of comets is believed to exist on the dim outer edges of the Solar System, about a light year from the Sun. The gravitational influence of passing stars nudges comets from this cloud into new orbits that bring them towards the Sun, where they become visible to us as ghostly, glowing apparitions.

When far from the Sun, a comet shines only by reflecting sunlight. At that stage it is small – only a few kilometres across – and faint. As a comet approaches the Sun it warms up, turning the ice into gas. Under the influence of the Sun's radiance the gases of the comet begin to fluoresce, in similar fashion to the gas in a neon tube, thereby considerably increasing the comet's brightness. Gas and dust released from the warming-up comet produce a halo or *coma* 100,000 km or so in diameter. At the centre of the coma is the nucleus, only a few kilometres in diameter and the only solid part of the comet, consisting of chunks of ice and rock. It would take a thousand million comets to equal the mass of the Earth.

Not all comets have tails, but many do. One part of the tail consists of gas blown away from the comet's head by the solar wind of atomic particles streaming from the Sun. The other part of the tail is made up of dust particles liberated from the head by the evaporating gases. Comet tails always point away from the Sun. A comet's tail can extend for 100 million km or more, yet despite its glorious appearance it is less dense than a laboratory vacuum. The tail of a comet gives it the appearance of rapid motion across the sky, but actually its motion against the stars is only noticeable by comparing its position from night to night.

A dozen or more comets may be visible through a telescope each year, although only occasionally do any of them become bright enough to be prominent to the naked eye. The comets that are seen each year are a mixture of known comets returning to the Sun and completely new discoveries. About a thousand comets have well-known orbits, and more are being discovered all the time. Dedicated amateur astronomers sweep the skies to discover new comets; each new comet is given its discoverer's name.

Many comets that stray into the inner regions of the Solar System have their orbits altered by the gravity of the planets so that they never again recede far from the Sun. The comet of shortest known period is Encke's Comet, which orbits the Sun every 3.3 years. It is so old that it has lost most of its gas and dust and is too faint to see with the naked eye.

The most famous comet of all is Halley's Comet, named after the English astronomer Edmond Halley who calculated its orbit in 1705. Halley's Comet reappears every 76 years or so, appearing this century in 1910 and 1985–6. Its orbit takes it from 0.6 astronomical units from the Sun (between the orbits of Mercury and Venus) out to 35 astronomical units (beyond Neptune and Pluto).

The dust lost from a comet disperses into space. The Earth and other planets are continually sweeping up cometary dust. When a particle of cometary dust comes whizzing into the atmosphere it burns up by friction at a height of about 100 km, producing a sudden streak of light known as a shooting star or meteor. The whole event is over in less than a second. On any clear night, a random five or six meteors per hour, visible as particles of dust, dash to their death in the atmosphere.

Halley's Comet pursues an elliptical path around the Sun, steeply inclined to the orbits of the planets. Every 76 years or so it moves from between the orbits of Venus and Mercury out to a greater distance than Neptune. *Wil Tirion.*

Brilliant Comet West spread its fan-like tail across the morning sky in March 1976. Note the difference in appearance between the broad, diffuse tail of dust particles (left) and the straighter, brighter gas tail. *Lowell Observatory*.

Such random meteors are termed *sporadic*. Occasionally, though, the Earth crosses the orbit of a comet and encounters a dense swarm of particles. This gives rise to a so-called meteor shower, in which meteors may be visible coming from one direction in the sky at the rate of dozens per hour. The area of sky from which the meteors seem to come is known as the *radiant*.

A meteor shower is named for the constellation in which the radiant lies. For instance the Perseids, an abundant shower of bright meteors

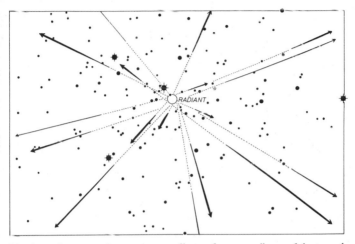

Members of a meteor shower appear to diverge from a small area of sky termed the radiant. This diagram shows the radiant of the Lyrid meteors, which lies in the constellation Lyra, near the bright star Vega. *Wil Tirion.*

which the Earth encounters each August, seem to radiate from Perseus; the Geminids from Gemini; and so on. One historical oddity concerns the Quadrantids, which come from an area in Boötes that was once part of the now-defunct constellation of Quadrans, the quadrant.

The strength of a meteor shower is measured by its zenithal hourly rate (ZHR), which is the number of meteors that an individual observer would see if the radiant were directly overhead. Since the radiant is seldom, if ever, at the zenith, the number of meteors actually seen per hour will be less than the theoretical ZHR. In addition, bright Moonlight will wash out the fainter meteors, again reducing the observed ZHR.

Amateur astronomers make valuable observations of meteor showers, counting the number of meteors visible and estimating their brightness.

Major Meteor Showers

Shower	Limits of activity	Maximum	Maximum rate (ZHR)
Quadrantids	January 1–6	January 3–4	100
Lyrids	April 19–25	April 21–22	12
Eta Aquarids	May 1–8	May 5–6	40
Delta Aquarids	July 15–August 15	July 28–29	20
Perseids	July 25–August 18	August 12	60
Orionids	October 16–26	October 21	20
Taurids	October 20–November 30	November 3	12
Leonids	November 15–19	November 17–18	10
Geminids	December 7–15	December 13–14	60

Typical meteors are of magnitude 2 or 3, but the most spectacular are brighter than the brightest stars, and the occasional brilliant meteor, termed a fireball, can cast shadows. Some meteors seem to split up as they fall, and some leave trains of glowing gas which fade in a few seconds. The table shows the main meteor showers visible each year. The ZHR is only a guide, and can vary considerably from year to year.

Asteroids and Meteorites

Between Mars and Jupiter orbits a belt of rubble known as the asteroids or minor planets. The asteroids were unknown until 1801 when the Italian astronomer Giuseppe Piazzi discovered Ceres, the largest of them, although astronomers had previously speculated that an unknown planet might exist in the suspiciously wide gap that separates Mars and Jupiter.

However, even if every asteroid was rolled together (about 2000 of them are known, and there must be thousands of others too small to be seen from Earth) they would make a body under half the size of the Moon. The asteroids are not, as has been suggested, the remnants of a former planet that was disrupted; they are merely the leftovers from the formation of the main planets.

Ceres itself is 1000 km in diameter and moves around the Sun every 4.6 years. Although Ceres is the largest asteroid it is made of a dark rock and is not the brightest asteroid. That honour goes to Vesta, which is composed of paler rock; at times it is just visible to the naked eye. Vesta is the third largest asteroid, 500 km in diameter. The second largest asteroid is Pallas, diameter 550 km. These asteroids, and several others, are bright enough to be followed in binoculars.

Ninety-five per cent of asteroids orbit in the main belt between Mars and Jupiter, but there are some notable exceptions. One group of asteroids, known as the Trojans, moves along the same orbit as Jupiter. Most significant from our point of view are the so-called Earth-crossing asteroids which have orbits that take them across the path of the Earth; these are also known as Apollo asteroids after the first of their type, discovered in 1932. Some of these objects may be the nuclei of extinct

Meteor Crater: impact of a giant meteorite in the Arizona Desert 25,000 years ago gouged out this massive scar, over 1 km across and 200 m deep. The iron meteorite was vapourized in the heat of the impact. *Ian Ridpath.*

comets. Famous Apollo asteroids include Hermes and Icarus, both of which have made close approaches to the Earth. Asteroids of this type must have hit the Earth in the past, and others may do so in future with devastating effects.

Objects that do reach the surface of the Earth are termed meteorites. One particularly large meteorite, estimated to have weighed a quarter of a million tons, crashed into the Arizona Desert about 25,000 years ago, digging out the now-famous Meteor Crater, 1.2 km in diameter. (Strictly, Meteor Crater is a misnomer; to be accurate it should be named Meteorite Crater, but the old name is too well entrenched to change now.) Most of the meteorite was destroyed on impact, but enough fragments remain scattered around the crater to show that the meteorite was made of iron.

In June 1908 a rather different kind of meteorite blazed into the Earth's atmosphere over Siberia, shattering at a height of 8 km with a force that caused devastation for up to 30 km around. This object is believed to have been a fragment from the head of Encke's Comet. Because of its low density it was completely disrupted before it could reach the ground to form a crater.

Several hundred meteorites land on Earth each year, but only about a dozen are actually picked up – the rest fall in uninhabited areas or into the sea. Most meteorites observed to fall are of the stony kind, and if they are not picked up immediately they soon weather away – at least, they do under normal conditions. But scientists exploring in Antarctica have now found large numbers of ancient meteorites, including many rare types, preserved in pristine condition by the natural deepfreeze of the ice cap.

Stony meteorites are usually termed chondrites, because they contain mineral-rich blobs known as chondrules. The small percentage of stones without chondrules are called achondrites. The most interesting class of stony meteorites are the carbonaceous chondrites; these have a high carbon content, and are believed to be among the most primitive rocks in existence, virtually unchanged since the formation of the Solar System. Most meteorites are believed to be chips off asteroids, but the carbonaceous chondrites may be fragments from the heads of comets.

A small intermediate group of meteorites is the stony-irons, which consist of about 50 per cent iron and 50 per cent rock. But the other main group of meteorites apart from the chondrites is the irons. These contain 90 per cent iron and about 10 per cent nickel. The largest known meteorite is iron, weighing about 60,000 kg (60 tons). It must have fallen to Earth quite gently, for it did not dig a crater and did not break up. It lies where it fell in ancient times, near Grootfontein in Namibia. By contrast, the largest stony meteorite weighs about 1770 kg (1.7 tons), and is one of a group of meteorites that fell near the city of Jilin, in China, in 1976.

Astronomical Instruments and Observing

Binoculars and telescopes serve two main purposes: they collect more light than the human eye, and they magnify objects. In astronomy, the first of these two aspects is paramount. Astronomers seek large telescopes not because they magnify more, but because their increased light-gathering power brings fainter objects into view and allows finer details to be distinguished.

Telescopes come in two main types: refractors, which use a main lens (known as the object glass or objective) to collect light, and reflectors, which collect light with a mirror. Astronomical telescopes have eyepieces which can be changed to give different magnifications. Binoculars are a modified form of refractor, in which the light path is folded

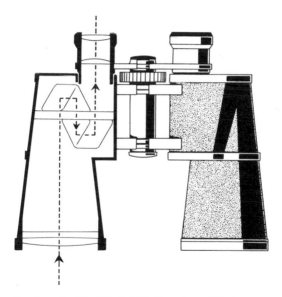

In binoculars the path of the light is folded by prisms. *Wil Tirion.*

by prisms to make them more compact. Telescopes that combine lenses and mirrors are becoming increasingly popular; these are known as catadioptric systems, and are best thought of as a modified form of reflector.

The first optical equipment that most astronomers own are binoculars. Indeed, they are virtually indispensable for any observer, even those who own a substantial telescope, for a binocular sweep can pick up objects for the telescope to home in on. Binoculars bear markings such as the following: 8×30, 8×40, 7×50, 10×50. In each case, the first figure gives the magnification, and the second figure is the aperture (in millimetres) of the front lens.

Any of the above-mentioned combinations of magnification and aperture are suitable for astronomy. Binoculars with powers greater than about $\times 10$ are difficult to hold steady; the image jumps around with each tremble of the hand, and they need to be mounted on a stand. Remember also that the higher the magnifying power for a given aperture, the fainter the image and the narrower the field of view. Binoculars of modest power give breathtaking wide-angle views of the heavens that telescopes cannot match.

Binoculars have the advantage that they are relatively inexpensive. By contrast, a small telescope – one with an aperture of 50–60 mm (2–2.4 inches) – can cost several times more than a pair of binoculars, with little gain in light grasp. What telescopes do offer is higher magnification and a tripod mounting to support the instrument. Unfortunately, in many mass-produced telescopes the mounting is not as steady as might be wished, and the image frequently vibrates for several seconds each time the telescope is moved. As a selling point, some small telescopes offer eyepieces with impressively high powers of over $\times 200$. But such high magnifications applied to a small telescope produce images so faint that little can be seen and are therefore a complete waste. As a good general rule, the maximum usable magnification on a telescope is $\times 20$ for each 10 mm of aperture ($\times 50$ per inch). These pitfalls

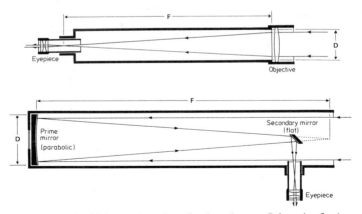

Top: The path of light rays through a refracting telescope. *Below:* A reflecting telescope of Newtonian design, the most common type used by amateurs. D is the telescope's aperture, F the focal length. *Wil Tirion.*

aside, there are many serviceable small telescopes available, which will bring into view a wide selection of the celestial sights mentioned in this book.

For more ambitious observers, a refractor of 75 mm (3 inches) aperture is the minimum required to begin serious work. Telescopes greater than 75 mm in aperture are usually reflectors, because size for size they are cheaper to make than refractors. Popular sizes are reflectors of 150 mm (6 inch) and 220 mm (8½ inch) aperture, which should bring into view virtually all the targets for observation listed in this book. As for cost, remember that a telescope is a precision optical instrument, and so you must expect to pay as much as you would for any similar instrument such as a good camera.

Reflectors used by amateurs are usually built to the design invented in 1668 by Isaac Newton. In the Newtonian reflector, light collected by the concave main mirror is bounced back up the tube to a smaller secondary mirror which diverts it into an eyepiece at the side of the tube. Inevitably, the secondary mirror blocks some of the incoming light from the main mirror, but this shadowing effect of the secondary is not significant and does not adversely affect the image. Another reflector design is the Cassegrain, in which the light is reflected back from the secondary through a hole bored in the main mirror. Large professional telescopes frequently use the Cassegrain design.

In the catadioptric systems that combine lenses and mirrors, the incoming light passes through a thin glass plate at the front of the telescope before falling onto the main mirror and being reflected as in a Cassegrain. The advantage of this design is that the overall length of the telescope can be made very much shorter than in a conventional reflector. The consequent saving in weight and space, allied with increased portability, compensates for the increased cost of a catadioptric telescope.

A telescope requires a mounting on which it can be supported and swivelled to point at various places in the sky. The quality of the mounting is just as important as the quality of the optics, for even the best telescope cannot be expected to show much if it is shaking around all the time, or if you have difficulty steering it to track objects as the Earth rotates. Sturdiness and smoothness of movement are all-important in a telescope mounting.

The simplest form of mounting is the *altazimuth*. This has two axes, one horizontal and one vertical, which allow the telescope to be pivoted up and down (in altitude) and to pan from side to side (in azimuth). In an altazimuth mount, the telescope is usually held in a fork on top of a tripod. The very smallest refractors have tripods so short that they have to be placed on a table. Such instruments are little more than toys and they are quite unsuitable for astronomy, where it must be remembered that the observer is looking upwards most of the time. A number of

An altazimuth mounting, the simplest form of support for a telescope. In an altazimuth mount, the telescope is held in a cradle which allows it to rock up and down and swivel from side to side. *Wil Tirion.*

refractors get over this problem by providing prisms known as star diagonals to fit at the eyepiece end. These bend the light so that the observer can look down into the eyepiece, which is certainly more convenient. In the case of Newtonian reflectors, tall tripods are not necessary since the eyepiece is near the top of the tube and so is normally quite accessible from a standing position.

A convenient addition to an altazimuth mount are small knobs, known as slow motion controls, which can be turned to move the telescope slightly in each axis. This is useful both for centring an object in the field of view, and for tracking it as the Earth spins. When high powers are in use, the Earth's rotation can carry an object out of the telescope's field of view in a remarkably short time.

A popular version of the altazimuth for large Newtonian reflectors is the Dobsonian mount, named after the American amateur John Dobson who devised it. In a Dobsonian, the telescope tube is made lightweight so that its centre of gravity lies towards the mirror end. The tube is supported at its lower (mirror) end by a wooden box with a Formica base that swings around in azimuth on pads made of Teflon (the slippery plastic material used in non-stick frying pans). Formica glides smoothly on Teflon, making a simple yet steady bearing. The tube pivots up and down in altitude on a shaft supported by more Teflon pads. Despite its deceptive simplicity, the Dobsonian has proved a very effective mounting. Cheapness and portability have ensured its widespread adoption.

Best of all for tracking objects is an *equatorial* mount. This has a main axis that is aligned parallel to the Earth's axis. This main axis is known as the polar axis because it points to the north celestial pole. It governs the telescope's movement in right ascension. The telescope pivots in declination on a second axis at right angles to the first, called (logically enough) the declination axis. Equatorial mounts are usually fitted with a small motor that slowly turns the equatorial axis at exactly the same speed as the Earth rotates. Once the telescope is pointed at a celestial object and the drive motor is running, the object will remain fixed firmly in the field of view for as long as the observer wishes. Experience soon demonstrates the desirability of a stationary image if you are trying to split a faint double star or to make a drawing of a planet.

A few words need to be said about what you can expect to see with telescopes of different sizes, and why. Anyone looking through an astronomical telescope for the first time is usually surprised to find that the image is upside down. There is a simple practical reason for this: to turn the image the right way up, an extra lens would have to be inserted in the eyepiece. Every time light passes through a lens (or is reflected off a mirror) some light is lost. Loss of light in an astronomical telescope

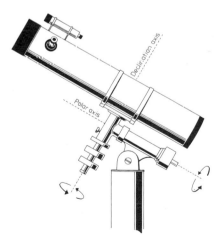

In an equatorial mounting, one axis (termed the polar axis) is set up so that it is parallel to the Earth's axis and hence points towards the celestial pole; the other axis (the declination axis) is at right angles to it. *Wil Tirion.*

The first of the great telescopes: the 2.5 m (100 inch) reflector on Mount Wilson, opened in 1917 and still working. With this telescope the existence of other galaxies and the expansion of the Universe was discovered. *Ian Ridpath*.

is undesirable, and in any case the additional lens is unnecessary. For astronomical purposes it does not really matter which way up the image appears, so the extra lens is left out, and the image remains inverted. Some telescopes come equipped with so-called terrestrial eyepieces which turn the image the right way up for everyday viewing.

When you make an observation you should record the date, the time, the instrument, the magnification, and the viewing conditions. Unlike binoculars and spyglass telescopes which have fixed eyepieces, the eyepieces in astronomical telescopes are interchangeable, offering a range of magnifications to suit the object under study. For instance, a star cluster or galaxy will require low magnification; planets are best viewed with medium magnification; and to split a close double star you will need your highest powers.

The magnification of an eyepiece depends both on its focal length and the focal length of the telescope. To work out the magnification an eyepiece will give in a particular telescope requires a little straightforward arithmetic. You simply divide the focal length of the telescope by the focal length of the eyepiece. The result tells you the power of that eyepiece. It is easy to see that the shorter the focal length, the higher the magnification. Eyepieces have their focal length marked on them.

Manufacturers often describe their telescopes in terms of focal ratio, such as f/6 or f/8. The focal ratio is the focal length of the lens or mirror divided by its aperture. If you don't know the focal length of your telescope you can easily find out by multiplying its aperture by the focal ratio. For instance, a 100 mm telescope with a focal ratio of f/6 has a focal length of 600 mm; if its focal ratio is f/8, its focal length is longer, at 800 mm. In the case of a 150 mm telescope of f/6 or f/8, the focal lengths would be 900 mm and 1200 mm respectively.

Now, assume that you have an eyepiece of focal length 20 mm. In a telescope of 600 mm focal length it will give a magnification of 600 divided by 20, which is 30 times. That is quite a low power for astronomical purposes. In a telescope of 1200 mm focal length, the same eyepiece will give a magnification of $\times 60$. An eyepiece of half the focal length, 10 mm, will give twice the magnification. Note that the aperture of the telescope is irrelevant in these calculations; the focal length is the sole figure governing magnification.

Where the telescope's aperture becomes crucial is in the matters of light grasp and resolution of detail. All other things being equal, a larger aperture shows fainter stars and finer detail than a smaller aperture, but exactly how faint the stars are and how fine the detail is depends on atmospheric conditions, optical quality and the observer's eyesight. Practical experience shows that the faintest stars likely to be visible through amateur telescopes of various apertures are as follows:

aperture	limiting magnitude
50 mm (2 inches)	11.2
60 mm (2.4 inches)	11.6
75 mm (3 inches)	12.1
100 mm (4 inches)	12.7
150 mm (6 inches)	13.6
220 mm (8.5 inches)	14.4

Incidentally, your eyes need time to become accustomed to the dark before you can hope to see the faintest objects; when you go out at night from a bright room allow at least 10 minutes for your eyes to become dark adapted. One useful trick when trying to glimpse faint objects is to use *averted vision* – that is, to look to the side of the object under study so that its light falls on the outer, more sensitive part of the retina.

The resolution, or resolving power, of a telescope is expressed in seconds (″) of arc. One second of arc is a very small quantity, equivalent to the size that an average coin would appear from several kilometres away. A telescope's resolution governs the detail that one can see on the Moon or planets, and the closeness of double stars that can be separated. Here are the theoretical limits of resolution for telescopes of various apertures:

aperture	limit of resolution (seconds of arc)
50 mm (2 inches)	2.3
60 mm (2.4 inches)	1.9
75 mm (3 inches)	1.5
100 mm (4 inches)	1.1
150 mm (6 inches)	0.8
220 mm (8.5 inches)	0.5

Under exceptional conditions, the limits of magnitude and resolution tabulated above may be surpassed; but in many cases, particularly in cities, they will not be achieved.

Finally we come to the atmosphere itself. Professional observatories are sited on high mountain tops to get above as much of the atmosphere as possible, but most amateurs are stuck with conditions in their own back yard, all too often including urban haze and glare from streetlights. There are two aspects of the atmosphere to take into account: its clarity, and its steadiness. A good index of atmospheric clarity is the magnitude of the faintest stars that can be seen overhead by the naked eye. The zenithal limiting magnitude should always be noted when meteor observing, because it affects the hourly rates. Atmospheric clarity needs to be at its best when looking for faint objects, particularly nebulae, galaxies and comets.

On the other hand, the steadiness of the atmosphere, known as the seeing, is paramount for planetary and double star observation. Turbulence in the atmosphere produces a boiling effect of the telescopic image, drastically reducing the resolution. Hot air rising from neighbours' houses can be a particularly annoying localized form of turbulence. Ironically, on a crystal clear night after a rainstorm the seeing can be particularly atrocious, whereas slightly misty nights, when the clarity is low, can be the steadiest.

Seeing is estimated on a five point scale: 1, perfect; 2, good; 3, moderate; 4, poor; 5, very bad. If a close double star cannot be clearly split on a night of indifferent seeing, examine it again on a better night.

Astrophotography

A camera can be used as an astronomical instrument. Photographs of the sky can be taken with any camera capable of time exposures. That includes most types of camera, other than the push button Instamatic or Polaroid variety. Depending on the length of the exposure, it is possible to photograph constellations, graceful curving star trails, groupings of planets and perhaps the occasional meteor.

Firstly, the camera must be loaded with fast film. Whether it is colour or black and white doesn't matter. Black and white has the advantage that it is more economical and can be easily processed at home, but colour transparencies are the most convenient for beginners. Colour negative film is not so good, as it requires special printing. Very fast colour slide films, rated ASA 1000, are available and these can produce beautiful results.

It is important to use the widest aperture on the camera lens. Camera apertures are quoted in terms of f/stops. Most camera lenses will open up to f/2.8, and many even wider, up to f/1.4. Cameras often have interchangeable lenses. A wide-angle lens will photograph more sky than a standard lens, but the images will be smaller and the constellation shapes will appear distorted towards the edge. Finally, the camera focus must be set to infinity.

A useful accessory is a cable release, which enables the shutter to open and close without shaking the camera. For time exposure, the camera needs to be on the setting marked 'B' (this, incidentally, stands for 'bulb', a legacy of the days when cameras were operated by air bulbs). On the B setting, the button must remain depressed to keep the shutter open, so the cable release must be locked for the length of the time exposure. Some cameras have a 'T' setting (for 'time'); on this, one press of the button opens the shutter, and a second press closes it.

The camera must be firmly mounted for the duration of the exposure. If there is no suitable tripod, the camera can be wedged firmly on the ground by pieces of wood, bricks, or stones. One good way of avoiding camera shake is to hold a piece of card over the lens while the shutter is opened. Once all vibrations have died out, the card can be moved away. The card must be put back in front of the lens before the shutter is closed at the end of the exposure. When the camera is pointing in the rough direction of the area to be photographed (stars are so faint they probably won't be visible in the viewfinder) you are ready to shoot.

The exposure used depends on what you are trying to achieve, and on the darkness of the sky. A good way to start is by photographing star trails, by leaving the shutter open for a length of time. As the stars move across the sky with the rotation of the Earth, they leave trails of light on the film. When you point the camera in the direction of the pole, the star trails will come out noticeably curved. If an exposure of the equatorial region of the sky is taken, the star trails will appear straight. Colour film will register an attractive range of star colours.

In an area with lots of streetlights, you should restrict the exposure to five minutes, or else the sky brightness will build up on the film and fog the image. Under darker skies, exposures of half an hour or longer will be possible without fogging. The only way to find out for sure is by experimenting.

If the exposure is kept short – 15–20 seconds – the Earth will not have rotated far enough to produce noticeable trails, and the constellations will be seen as they appear to the naked eye. Stars down to 6th magnitude will be recorded. Photographs taken with short exposures in twilight, or with moonlight illuminating the surrounding landscape, can be particularly attractive. From time to time, several planets form an interesting grouping which is worth capturing with a short exposure. Always take a succession of photographs with different exposures to be sure of getting the best result. Use of a telephoto lens can sometimes add drama to a scene – for instance, a partially eclipsed Moon rising over a distant skyline. But you must remember that the effect of the Earth's rotation shows up more quickly through a telephoto, so exposures should be restricted to no more than a few seconds to avoid trailing of the images.

When a bright meteor shower such as the Perseids or Geminids is due, try a series of long exposures with a wide-angle lens in the hope of catching some. Meteors dart across the sky so quickly that only the brightest will register on your film, and many of them will in any case fall outside the camera's field of view. But the excitement of photographing the brilliant streak of a meteor more than compensates for the disappointment of the other frames that show nothing but star trails.

If a camera is placed on an equatorial mount it can be guided to prevent star trails forming. Guided exposures of only a few minutes' duration will capture stars far fainter than those visible to the naked eye. Most exciting of all is to use a camera to photograph what can be seen through a telescope. A single lens reflex camera is essential for this. The camera lens must be removed completely and the camera body attached to the eyepiece mount (adaptors are available for this). In effect, the entire telescope is acting like a super telephoto lens.

With such a system craters on the Moon, the belts of Jupiter and the rings of Saturn can be photographed. Exposures range from a fraction of a second for the Moon to a second or more for the planets. Even longer exposures, of several minutes' duration, can reveal delicate details of faint nebulae and galaxies. But that is only for advanced astrophotographers, and the techniques involved are outside the scope of a basic book such as this.

Index

All named first and second magnitude stars are included in the index, e.g. Adhara. Stars such as η (eta) Carinae will be found in the main body of the text under the relevant constellations. Page numbers in bold type denote photographs or diagrams.

Biographical Notes

Ian Ridpath is an English amateur astronomer, and an author and broadcaster on astronomy and space. Among his other books are the *Encyclopedia of Astronomy and Space*, and the *Longman Illustrated Dictionary of Astronomy and Astronautics*. Wil Tirion is a Dutch amateur astronomer and graphic artist. In 1981 he published *Sky Atlas 2000.0* which was immediately hailed as the finest atlas of its kind. Ian Ridpath and Wil Tirion are also authors of the *Collins Gem Guide to the Night Sky* (published 1985).